River Time

MARY A. HOOD

RiverTime

ECOTRAVEL ON THE WORLD'S RIVERS

STATE UNIVERSITY OF NEW YORK PRESS

Published by
STATE UNIVERSITY OF NEW YORK PRESS, ALBANY

© 2008 State University of New York

For information, contact State University of New York Press, Albany, NY
www.sunypress.edu

Production and book design, Laurie Searl
Marketing, Susan M. Petrie

Library of Congress Cataloging-in-Publication Data

Hood, Mary A., 1944–
    Rivertime : ecotravel on the world's rivers / Mary A. Hood.
        p. cm.
    Includes bibliographical references and index.
    ISBN 978-0-7914-7389-4 (hardcover : alk. paper)
        1. Rivers.   2. Natural history.   3. Hood, Mary A., 1944– Travel.
    I. Title.

QH97.H66 2008
577.6'4—dc22   2007034184

10 9 8 7 6 5 4 3 2 1

Contents

PART III. OTHER GREAT WORLD RIVERS

Illustrations

Introduction

American literature is full of rivers. Indeed, the world's literature is full of rivers, and rivers are more than just background; they are an essential element of our stories. Our most ancient stories are deeply entangled in rivers: the River Styx as the gateway to the underworld, Moses left in the reeds on the Nile, the Israelites crossing the River Jordan, and the baptism of Jesus occurring in that same river. Scheherazade's tales of *1001 Arabian Nights* were probably told on the banks of the Euphrates and Tigris, as were the ancient legends of Gilgamesh. That our earliest civilizations emerged on rivers reflects a broader truth: rivers spawn civilization. Because of our deep roots in rivers, because rivers provided sustenance and facilitated travel, they are one of our strongest connections to and precious elements of place.

The enormous body of nonfiction on rivers reinforces the notion that they serve as important, defining places. It would be impossible to mention all the important writings on rivers, but a few recent American publications illustrate the point that we are deeply concerned and involved with our rivers. *The Rivers of America: A Descriptive Bibliography* (Fitzgerald 2001) describes the history behind many of the country's rivers. The writings of environmentalists, conservationists, and geoscientists (Mancall 1996; Palmer 2001; Postel and Richter 2003; Wohl 2004) focus on the vital links rivers make. American nature writers such as Gary Snyder (1994, 1995), Barry Lopez (Murray 1998), and Wendell Berry (1997) define our connections to the earth in terms of rivers. They insist that rivers define who we are on the most primal level. These writers also remind us that a river's watershed may be the quintessential expression of bioregionalism.

Over the past four years, I have traveled to many places. All these ventures entailed river travel. I never initially planned my journeys this way, and I was unaware of this connection until I began to reflect on these travels. My river time has included the six great world rivers as well as many small rivers in the United States. Even where rivers were absent, they had often sculptured the landscape before disappearing. The rivers of Three Corners (Alabama, Florida, and Georgia) have wrought such effects. Even in the driest places, as we will soon see, rivers leave their marks. As I traveled about my country and other parts of the world, I could not help but be aware of rivers' omnipresence.

By now I recognize that rivers are stories. Each has a beginning, a middle, and an end. It would take a lifetime to listen to the story of even a single river, so the stories told here are mere vignettes. They include a bit of natural history, and they explore some environmental issues, while also reflecting my thoughts about life. Like our stories, rivers connect us. They connect us to our origins, to our biological heritage, to the natural world, and to each other. Even as rivers change and are changed, their universal truth remains. They are the stories that engage us in all the ways our best stories do.

The chapters in the middle of the book are about the world's largest river, the Amazon. This river occupies the center of the book because it is the most important river on earth; it is the essential force regulating the earth's biological systems. These chapters look at some of the Amazon's many biological wonders, from bats to bacteria, from birds to botos, always with the river's companion trees as part of the story. The remaining world rivers offer other themes. The Mississippi River chapter is about the Atchafalaya delta and the land changes created by both natural and human forces. The Tizsa River, a tributary of the Danube, borders a preserve that functions not only as a wildlife refuge but also as a base for agriculture and aquaculture. The Yangtze is an ancient waterway undergoing the transformation of economic development. The Ganges, yet another old-world river, illustrates the paradox of human reverence for rivers along with their devastation by the overwhelming human populations that live along them. Finally, the chapter on the Nile, the oldest river to support a civilization, is about the river as a tourist attraction.

The chapters on smaller United States rivers explore more narrowly focused environmental topics. Three Rivers, New Mexico, addresses the need to write our world. The rivers of Three Corners, formed during the Pleistocene epoch, describe a unique watershed known as "the canyon lands of Florida." The Willamette recalls an encounter with the endangered northern spotted owl; the Penobscot recounts the great north woods of Maine; the Flathead explores the fires (and corporate control) of the west. The Klamath describes the great redwoods; the Conhocton focuses on small New York towns; the Ellijay illustrates how making applesauce honors both the land and its rivers. The Tensaw depicts the rare bigleaf magnolia, and the Cumberland expands the topic of magnolias and includes other rare and unusual plant species of the Appalachian forest. The Alligator River describes the red wolf and tundra swans in tidewater North Carolina. Finally, the Yellow River traces some of the conservation, preservation, and restoration efforts that help keep rivers natural and healthy.

Over a period of four years, I traveled through the cycles of the seasons. Each season brought its distinctive colors to these stories, whose twists and turns provided glimpses of rare plants, endangered species, and even wildfires. The rivers offered exotic terrains, to be sure, but also familiar landscapes such as farms and small towns. Some rivers showed how we humans can devastate the places we inhabit. All the rivers I visited point to what human beings experience on twenty-first century earth. Here are the stories the rivers hold and my reflections on river time.

PART I

SMALL UNITED STATES RIVERS

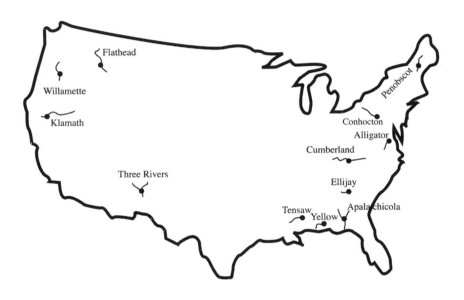

FIGURE 1. Map of small rivers of the United States

Three Rivers:
Writing Our World

In the Tularosa Basin of New Mexico, three dry rivers converge near an old basaltic ridge. Nearby, fifty acres of rock art adorn the site, and a state park called Three Rivers Petroglyphs Site preserves the ancient stone writings. In all directions, unusual geological formations spread out to form a landscape of contrast. To the north, black lava fields called the Valley of Fires, or the Malpais ("badlands"), make up a plain that looks like plowed earth. To the south, the smooth white gypsum dunes of White Sands National Monument offer a stark reversal. To the east, the Sacramento and White Mountains, and to the west, the San Andres

Mountains form the rim of the basin, which boasts one of the most beautiful and inscrutable landscapes of the southwestern United States. Half a million years ago, the basin was a shallow inland sea fed by many rivers. The rivers' depositional sediments formed a buildup of limestone, shale, gypsum, and sandstone. The sea was pushed upward like the crust of a rising loaf of bread by tectonic action, and when it sank in the middle, it created a 60-mile wide and 150-mile long depression. Water from the mountains flowed into the depression but had no way of escaping except by evaporation. At the end of the last ice age, about 10,000 years ago, the climate was wetter and there was water in the large lake. Today, the lake is almost dry and is what geographers call an alkali flat.

What makes the Three Rivers site such an intriguing gallery is its 21,000 petroglyphs. Images of plants, animals, insects, sky and earth objects, and abstract designs are chiseled into the rocks. They were carved a thousand years ago by the Jornada Mogollon culture, a people probably related to the ancient Mimbres. The three-quarter-mile long trail that wound through the rocks was well marked, and as my traveling companion and I climbed, we tried to match the petroglyphs on the stones with the pictures in the park brochure. There was a stick man, a fish, and a stick man inside a fish (Could this be a Jonah-in-the-belly-of-the-whale story?), a deer with arrows in its side, a turtle, a rattlesnake, a bighorn sheep, a lizard, and more. It seemed the people who inscribed these stones were determined to create a pictorial list of all the creatures and objects in their world. The abstract signs, squares, triangles, circles, coils, mazes, and ripples were equally vivid, but we could only guess what they represented.

As we moved through the boulders, I wondered why we humans make images that represent our world. Why do we write our world? When I was eight, I was fascinated by live oaks; my third-grade notebook was full of penciled trees. When I learned to write sentences, I wrote poems about live oaks. Nature writers offer many reasons for writing our world (the American Society for Literature and the Environment's bibliography gives a list of such nature writers [ASLE 2005]). Some claim that we write to "make place," to make sense of our world, or to bring order (Dobrin and Keller 2005); some (McKibben 2003) suggest that we write to defend it, while others (Dallmeyer 2004) suggest that it allows us to envision a better world. Solnit (2007, 1) proclaims "It was place that taught me to write." Phillips, in *The Truth of Ecology: Nature, Culture and Literature in America* (2003), explains that we simply need to describe our world. He distinguishes nature writing from natural history according to the involvement of the narrator. Though both are descriptive, nature writing has a more involved narrator than natural history. He claims that "nature writers . . . present themselves as both keen observers and as spiritual barometers sensitive to the pressures that weigh upon body and soul here on earth" (Phillips 2003, 187). For me, writing has been a way of honoring the world and a way to declare love. My childhood loves were names written on the white pages of my notebook. But that was only the beginning. The list increased, and over the years the words became more than people's names, more than places, more than trees. What I write today is still oaks and plants and birds and the natural world. I write to make connections, to make conversation, and with the hope that I might help others make connections. Was that what the petroglyphs were for? Are we as writers always trying to be part of a connection, part of a larger conversation? Did those ancient peoples need to write their world as much as I do?

From the highest point on the scripted rocks, White Sands was a hazy dust cloud on the horizon. The day before, we had trudged up the powdery dunes carrying plastic sleds and tobogganed down the calcium hills. For hours, we were kids on the slide. So different from the hard quartz sand of Florida beaches, this sand was like talcum powder. My toes became so dehydrated they looked like white raisins. It was only when I remembered that I had used this same mineral, calcium sulfate, as a desiccator in the laboratory that my dry, cracked feet made sense. Calcium sulfate takes the water out of everything.

Earlier that morning, we had hiked to the alkaline lake and walked among the fields of crystalline selenite. Here the calcium was hard crystalline gypsum. Bizarre amber shards protruded from the rock, creating a trail of broken glass. This arid land, with its calcium dust and calcium glass, its white talc and glassy outcroppings, was mineral chemistry at its most artistic.

The White Mountains that form the eastern lip of the basin contain a fifty-thousand-acre wilderness that is part of the National Wilderness System. A few days earlier, we had discovered a Lewis's woodpecker in Nogales Canyon feeding on elderberries. The dark bird was a stunning sight against a background of autumn mountain foliage. Back and forth the bird flew from the elderberry tree to the yellow oaks. Against a backdrop of pale auburn Rocky Mountain maples and a blue denim sky, with the high slopes covered in yellow aspens and dark evergreens, the bird was the centerpiece of a majestic canvas.

Watching the woodpecker, I thought about how much I loved birding. The recent rash of books about birders (Osborne 2003; Cocker 2003; Cashwell 2003; Obmascik 2004; Koeppel 2005, to cite only a few of the best-sellers) describes an activity that has become both popular and popular to write about. These books offer some of the reasons why people bird. They describe birders who travel all over the country (some all over the world) and birding as an eccentric hobby, a game, a competition, even a social event. In an article entitled "13 Ways of Looking at An Ivory-Billed Woodpecker," Hitt (2006, 43) explains that "the act of birding, ultimately, is an act of storytelling." For me, birding has always been a connection with the natural world, and with that connection has come an awareness and an understanding that we must keep what is left of the natural world and try to restore what has been destroyed. Maybe if I write well enough, maybe someone will be moved to do the political and economic things needed to keep our world as natural as possible. Maybe, like Scheherazade, my life depends on it.

Beyond the mountains, the high chaparral of Lincoln County rolls on like a Christmas holiday, with its red hills and green junipers. The day before, we had walked one of the evergreen-speckled canyons with Anne. Her ranch in Arabella was a gathering place to share a potluck dinner with the Lincoln County Bird Club. After a smorgasbord of tasty southwestern cuisine, Anne led us to an old ranch site and cantina. Along the way, Gene discovered a perfect arrowhead the size of a thumbnail. Pat spotted a horned toad lizard, and we watched it flatten its body against the rock like a shadow. When its camouflage didn't work and we were still there, the mini-dinosaur scampered off, its spiky head trembling in fierce retreat. Pat pointed out the amethyst shards of old bottles and told us how old glass contained minerals. The manganese, with age, turned the glass the color of plums. Anne showed us the remains of the old stone foundations and cellar. She told us the family had six kids and lived totally off the land. A goat pen revealed the probable source of milk and meat, and we speculated where the garden might have been and how water was brought up to the garden from the stream nearly a mile down the canyon. We marveled at the resourcefulness of people who could live in such a harsh land. And yet the history of all indigenous Native Americans was just that subsistence story.

To the north of Three Rivers, the dry flatland known as Jornada del Muerto ("the valley of death") stretched out to another hazy horizon. Here, Trinity Site marked the testing of the first atomic bomb in 1942. This was the place where my country (the only country to ever use the nuclear bomb) made weapons of mass destruction. To create a thing of such destruction, to use it to destroy other living creatures, even my country's enemies, was beyond my understanding. I would never be able to make order of that. It has been argued that the dropping of the bomb on Hiroshima saved countless American lives, but all those survival arguments of "it was either them or us" just don't make sense. I wondered if there were petroglyphs that stood for fear and madness.

The afternoon sun beat down from a cloudless sky, and the plains rippled with heat. We climbed down from the hot rocks and left the ancient graffiti of Three Rivers. Driving over the dry river beds, I imagined the people who had carved the petroglyphs leaving with perhaps the same thoughts I had when I finished writing. Maybe someone would read what I had written and be moved by it; maybe they would know that I had tried to evoke for them the beauty of the natural world. In that knowledge they might find an awareness of a grand and beautiful connection, an essential connection that rivers can make, even if they are dry.

Apalachicola and Other Rivers of Three Corners: Fossil Rivers

A weekend looking at plants might not seem very exciting, but that was how we members of the Florida Native Plant Society, Long Leaf Pine chapter, spent a few days one November in the Three Corners region of Florida, Georgia, and Alabama. The region is unique because ancient Pleistocene rivers molded the land and created a rare mixture of habitats: southern Appalachian hill forest, southern hardwood swamp, and sand hill pine barrens. Cutting through underlying limestone, the ancient rivers created ridges, ravines, and rolling hills. With the exposed limestone and deciduous hardwood forests, the terrain has the appearance of the Appalachian foothills. Yet in the floodplains, the lowlands are wet hollows typical of southern hardwood bottom-lands, and the drier uplands have all the characteristics of the piney woods.

Three state parks—Three Rivers, Torreya, and the Marianna Caverns—provide access to the habitats, and John Tobe, a state botanist and expert on the ecology of the region, provided the expertise. He led our band of curious botanizers on a merry exploration of this land sculptured by fossil rivers.

"Get in here, right now!" the lady hollered in an angry voice. The snibbling of a cranky kid as he crawled into the tent was muted by another yell. "Don't you . . ." and the rest of the scold was garbled. We plant society members dashed into the woods to get away from the shouts. Packed in like graham crackers, the campers were a nosy bunch, and the campground at Three Rivers State Park was buzzing with the comings and goings of pickup trucks, boats being docked, and people milling around. We found the trail and moved away from the loud gatherings as quickly as possible.

The Chattahoochee and Flint Rivers meet to form Lake Seminole, and from this lake, the Apalachicola River meanders down to the Gulf of Mexico. On the southeast shore of Lake Seminole, the Lakeside Trail creeps up piney hills, through hardwood forests, and loops back to the river front. The trail begins as a damp path along the river where dense shoreline vegetation blocks much of the river's view, but occasionally the path leads directly down to the water's edge. At these clearings the confluence of the rivers can be seen as the amoeboid lake. Supposedly, the lake supports good fishing for bass, blue gill, speckled perch, and bream. Formed when the Jim Woodruff Dam was completed in 1956, it is now a flooded river swamp. Although few fishermen were out on the lake, the sound of someone cranking a boat motor time and time again filled the air with an awful grating noise.

The trail follows the lake's edge for about a quarter of a mile and then veers off up a hill. Blood root is common, and banana spiders were busy hanging their nets like laundry between every other tree. A colony of mushrooms had settled on a fallen log, adding a touch of yellow to the trail, but there was not much else of interest. Not a single bird was about, the sky was overcast, and we were glad at the end of the afternoon to return to our lodging for the evening. Maybe it was the crowded campgrounds, maybe it was the weather, maybe the dam, but the river seemed oppressive and congested. There is something about a dam that takes the spirit out of a river; there is something about a crowd that takes the beauty out of a riverside.

The next morning we eager botanizers gathered at Torreya State Park. The directions to the park were a little confusing, so we wandered a bit among the back roads through hardwood and longleaf pine forests before we finally found it. Located on the Apalachicola River about twenty miles south of Marianna, Florida, the park is surrounded by land purchased over a thirty-year period by The Nature Conservancy. The six-thousand-acre Nature Conservancy preserve called the Apalachicola Bluffs and Ravine Preserve, described by Vaughan (2004), has an ecosystem of ravines called steepheads. With a touch of humor, Means (1991) dubs the region "Florida's canyon lands" because of these ravines and gullies.

Torreya State Park has several botanically interesting trails, including the Weeping Ridge Trail, which circles the park, and the Bluffs Trail, which loops from the high bluffs near an old colonial house down to the river's edge and back. We trekked off with John Tobe leading the way to explore the plants along both trails.

In a patch of sandy soil near the Weeping Ridge trailhead, we spotted an interesting herb called the purple milkwort (*Polygala polygama*). This thin-leafed perennial, also known as bitter or racemed milkwort, is found throughout the eastern United States in sandy wooded areas and prairies—and in the eastern but not the western part of the Florida Panhandle. What makes the plant unique is that it produces a flower below ground; botanists call this a cleistogamous flower (Kaul et al 2000). This characteristic is so rare that of all the 250,000 species of flowering plants, only thirty-six species produce such cryptic flowers. The very idea of an underground flower seems contradictory. After all, if the function of a flower is to be seen, to be showy, to attract attention so that some creature might come along and pollinate it, why hide in the ground?

As we observed the plant, I recalled an even rarer group of underground flowers, the underground orchids (Jones 1988). To date only two species of underground orchids, *Rhizanthella gardneri* and *R. slateri*, have been found and only in Australian mallees (eucalyptus-dominated shrub lands). These rare, down-under orchids spend their entire life cycle as subterraneans, buried beneath the sandy soils of the desolate scrublands. The orchid species, *R. gardneri*, the parasitic broombush, *Melaleuca uncinata*, and a mycorrhizal fungus form an interdependent threesome whose exact chemical relationships are unknown. The orchid is pollinated by gnats that crawl under the leaf litter and incidentally by termites and mosquitoes (yes, these annoying insects have a practical function beyond aggravating us and being disease vectors). Seeds produced in a small berry are dispersed by animals that eat the berries and defecate, transporting them to a new spot. When cut, the flowers smell like formalin. Almost nothing about the biology of the other orchid species, *R. slateri*, is known. Unlike the orchids, the milkwort is a more "normal-looking" plant. It has leaves and long stems and produces an above-ground flower that blooms in June. The purple aerial flowers smell a little like spice cake.

Observing the milkwort, I thought, if a flower produced a scent, even if it was a scent like formalin, as is the orchids', or spice cake in the case of milkwort, and if that scent attracted ground-burrowing insects, and those insects pollinated it, the flower had done what flowers were meant to do. How remarkable it is that plants have so many ways of doing what needs to be done. In the botanical world, it gives new meaning to the phrase, "there are many paths to the mountain."

The red clay trail sloped down through a longleaf pine-wiregrass habitat, and the morning sun reflected off pine needles and long blades of wiregrass. Everything shimmered. Every surface sparkled and spangled like a tinseled Christmas tree. A Bachman's sparrow flitted into a thicket, and pine warblers chattered high in the branches. Almost before we had time to realize that we were in a longleaf pine forest, the terrain suddenly changed and we were in a hilly deciduous forest. John pointed out some of the common trees: elm, ash, hickory, beech, and oaks formed the taller canopy with Florida maples (*Acer barbatum*) and chalk maples (*Acer leucoderme*) filling in the understory. With the leaves raining down like a New England autumn, John showed us the difference between the two maples. Chalk maples, he said, have droopier leaves with hairy undersides.

When the chalk maples' seeds dropped, they whirled to the ground like helicopter blades. John shook a small tree to illustrate the seed type (botanists call them samara) and they fluttered to the ground like whirligigs. Two of our troops dashed about trying to catch them. Stuffing them into their pockets, they exclaimed they were going to plant them when they got home and grow chalk maples. Not a bad idea. Trying to grow a non-indigenous variety, such as a New England maple in a Florida yard, is not a good idea because it usually will not do well. But planting a native species, such as a chalk maple (especially if limestone is available) or a Florida maple, is far more likely to work. It is a concept that should apply to all planting efforts. In other words, plant native.

Brown nuggets speckled the trail. The ground looked like a marble tournament with so many hickory nuts. There were mocker nuts, pig nuts, and bitternuts. When one of the group picked up the little shells, we teased her about bringing home treasures. We laughed when she admitted that she was going to use them as home decorations. I thought, how strong is this need to bring the beauties of the natural world back into our homes, and it does not stop when we outgrow Campfire Girls or stop thinking of nut shells as fairies' caps.

We descended onto the Apalachicola River floodplain and the terrain changed again. The trail was steep, a reminder of why the term *steepheads* has been used to describe these ravines. Spike moss, liverworts, and climbing hydrangea clung to the limestone banks; horsetail, southern shield, and Christmas ferns grew in the flatter spots. Tupelo, sweet gum, and cypress rose from the spongy soil, their flared trunks resembling a group of teenagers in bell-bottom jeans hanging out. Southern magnolia, the rare Ashe's magnolia (*Magnolia ashei*), chinkapin oak, shumard oak, black oak, and box elder made up the complex mosaic of other hardwoods.

Along the Bluffs Trail, John pointed out the Florida yew (*Taxus floridana*) and the Torreya tree (*Torreya taxifolia*), whose branches and needles look a bit like spruce. The rarest of the rare, the Torreya tree is found only in this small region along the banks of the Apalachicola River. It grows nowhere else on earth. The idea that some species have such narrow home places, while other species have such a wide range, made me wonder about what makes home. How is it that some plants, like the Asian climbing fern, well known as a troublesome invasive which we encountered all over the trail, can grow almost anywhere, while others, like the Torreya tree, can only grow in this one tiny spot in all the world?

Indigenous palms, such as the needle palm and the sabal palm, grew among their more common relative, the saw-tooth palmetto. A barred owl spooked and flew off towards the river, hooting her annoyance at being disturbed. Dutchman's pipe vine, the host plant for the pipevine swallowtail, tangled in the top of a thicket. Stinging nettle, host to the red admiral butterfly (with its lovely genus name, *Vanessa*), also grew in patches along the trail. When someone remarked that the red admiral circumnavigates the world, I paused to let that sink in. Circumnavigate? Could that be true? It was hard enough to imagine an artic tern circling the globe, but a delicate butterfly, well, that bordered on the miraculous.

Water oats and trillium filled in the shaded places where only a little sunlight was able to sneak through the dense branches. A parade of shade-loving plants, like sensitive fern, fox grape, a silver bell tree, butterweed, and others, lined up along an "abracadabra" creek. Water oozed up from the ground, meandered a little ways, and then vanished again into the ground like magic. The creek's milky blue waters, full of calcium, supported rare fish and salamanders, John said. Somehow it seems fitting that in this place of rare plants and animals, an aquamarine creek would suddenly appear, flow a short distance, and then just as suddenly disappear back into the limestone. Abracadabra!

We returned to the high bluffs overlooking the river and watched the syrupy waters of the Apalachicola bend around a lazy curve. After a time watching the river ooze its way south, we headed off to our next destination, the Marianna Caverns.

About thirty miles west of Three Rivers State Park, the Marianna Caverns also had a Bluffs Trail, and it ran along the east side of the Chipola River floodplain. The underlying limestone, known as karst topography, was full of lacy underground caverns, sinkholes, and pits, and the Bluffs Trail was a lime-lover's paradise. Columbines and zephyranthes known as atamasco lilies (*Zephyranthes atamasco*) covered the water-carved boulders that lined the trail. Atamasco lilies, or rain lilies, are not really lilies but amaryllises, and after spring rains they bloom into white stars. Our expert botanist, John Tobe, explained that in March the rocks turned into stunning wildflower gardens that looked like the hanging gardens of Babylon. He was so enchanted with the limestone rock gardens that he decided to make one at home by getting some "feral cement" and planting some of these wildflowers on the concrete.

The idea of "feral concrete" had never occurred to me. I think of concrete as an alien substance, a plant terminator, of sidewalks that mow down grass, of parking lots that wipe out meadows, of concrete buildings that displace forests. I think of concrete drains diverting billions of gallons of water from the soil to sewage treatment plants. To imagine concrete as something to assist plants was a new way of thinking. Of course, concrete is limestone, so why not imagine it in a new way, as a plant helper? I began to imagine cities where high-rises served as trellises, where freeways became long row gardens, where driveways became orchards. The idea was really appealing. If we could make concrete a life-giving substance rather than a life-destroying one, what a wonderful world it would be. And the term *concrete jungle* might turn into something entirely true.

We walked past crevices in the ground covered with metal bars. Openings to the bat caves where grey bats made their home, the pits looked like entrances to the underworld. The metal bars were there to keep out curiosity seekers and cavers. It was hard to imagine anyone wanting to enter such a hole, but apparently there are many people who do. Even after reading the lovely *Entering the Stone: On Caves and Feeling Through the Dark* (Hurd 2003) and learning that a lot of people explore caves with great relish, I found it difficult to imagine entering a cave. Hurd's writing is beautiful and enticing, but still, I could not understand someone actually enjoying the process of entering the darkness of a cave. I get shivers just thinking about the physical reality of it. If the cave serves as Hurd's metaphor for death, and if I had a choice of metaphors, I would choose not to enter the cave. I would, like Virginia Woolf, fill my pockets with stones and walk into the river.

As our day ended and we sat on the limestone blocks at the end of the trail, I wondered about the nature of fossil rivers. Those wild ancient rivers once cut through the land exposing the underlying limestone bedrock and, in doing so, created unique environments. The environments in turn supported unusual plants and, in some cases, plants so rare they could only be found in this one place on all the earth. Perhaps the legacy of some rivers is the creation of unusual landscapes, and with the creation of such habitats, the river gives the gift of rarity.

Willamette River:
Mousing the Owl

From the mountains of west-central Oregon, the Willamette River flows north between the Cascades and the Coast Range and empties into the Columbia River near Portland. It cuts a wide valley through the state, a valley that is the most productive agricultural land of the region and an important birding area (birders call this an IBA). The Willamette flows through the center of Eugene, Oregon, and it was there I first encountered the river. We were in Eugene for the American Birding Association (ABA) annual meeting, and every evening for a week we walked along the river paths.

One of the most popular ABA events was the northern spotted owl field trip. Sixty people lined up at 4:30 AM to board the
morning bus, and as soon as the door opened we clambered up the
dark steps to find seats. The bus lumbered off towards Goshen
with its load of sleepy birders. As the sky lightened, the bus turned
off the interstate. Many turns on back roads and hairpin curves
later, we were high in the coastal mountains. Forests of fir and
hemlock hugged the mountainsides as the bus climbed higher and
higher and the road became narrower and narrower. The first stop,
an old-growth forest of Douglas fir and western hemlock, was stunning. The giant evergreens were colossal columns in a cathedral,
but the canopy was higher than any ceiling. Maybe a domed sta-
dium would come closer to the dimensions of that canopy, but
certainly no human structure had the transcendent feeling of that
forest. Beams of light solid enough to be ramps streamed down
through the evergreens, and I imagined walking one of those white
staircases into the clouds.

A wildlife biologist with the U. S. Forest Service, Eric
Forsman, met the bus at the next stop and soon located two spotted
owls perched on the dark branches of a fir tree. With mutterings of
delight, our flock of eager birders quickly gathered at the edge of
the road and peered into the trees. Binoculars raised and scopes
adjusted, we stared at the birds while Forsman described some of
his studies stretching over a twenty-year period (Forsman et al.
1983, Forsman 2002). Probably more is known about this owl than
any other owl, he said. His current research efforts involved a
monitoring project designed to determine population numbers and
distribution. By banding and recapturing, he and his colleagues
could determine territorial range, reproductive success, and the
general stability of owl populations. The study site we were in had
thirty pairs of owls, but the total number of owls in the Pacific
Northwest probably numbered around 6,000.

The latest data, he explained, showed that barred owls were
proving to be almost as destructive to the spotted owls as humans

were. When barred owls moved into spotted owl territory, the spotted owls left. Whether barred owls chased away the spotted owls, or competition for food was too intense, or there was some other factor was not clear. He also mentioned that where barred and spotted owl territory overlapped, hybrid species were found. In other words, the two species were interbreeding. That sort of species interaction appealed to me. Blending of species, instead of competition to extinction, seems a gentler sort of biological accommodation.

A young woman who worked with Forsman carried a large plastic box. She set the box down near the trees where the owls perched. Until she lifted the lid and took out a mouse, it did not occur to me what the little cage was for. She placed the mouse on the end of a long stick and held the stick high in the air. Out of the darkness, the owl swooped down, grasped the mouse, and took it back to its tree. Three times she placed a mouse on the stick and held it up, three times the male owl swooped down and grabbed the mouse. He ate one and cached two. In the time it takes to swallow an aspirin, the owl had eaten the first mouse. Then the female was offered mice and she feasted.

Each time the owls were presented with a mouse, I watched closely, trying to see the details of the catch, but each time it happened so quickly, it was impossible to observe. One instant the mouse was sitting on the end of the pole; the next he was gone. One moment the owl was perched on his dark branch, and the next he had returned to his dark spot. With a swift rip and gulp the mouse was gone, and the owl stared back at us with black inscrutable eyes.

Later, as we rode to another site to see other birds, I wondered about what we had seen. Was it about biologists feeding mice to owls? Was it about birders seeing the endangered northern spotted owl, the bird that became the symbol of the struggle to save the old-growth forests of the Pacific Northwest? Was it about biologists saving the spotted owl? Was it PR by the Forest Service to gain support for their efforts to protect the spotted owl? It was probably all of those things, but mousing the owl, as one of the ABA organizers called it, made me a little uncomfortable. The idea that sixty people would pay $75 each to see owls fed mice did not seem like something we should be doing. It felt too much like a gladiator event. Yet I was fascinated with the owls and had spent every second watching them. Biologists have to capture and band the owls for their studies, but sometimes I wonder about us humans. We just cannot leave things alone. We destroy the owl's habitat. Then we study the owls in order to save them from extinction. We humans are always meddling in the lives of other species. We make so few accommodations for species other than our own, and then when faced with the realization that our behavior results in their destruction, we try to redeem ourselves using more manipulation.

In the wild, the spotted owl eats red tree voles. These arboreal rodents live all their lives in the canopy of old Douglas firs and spruce. Rarely touching the ground, they live secretive aerial nocturnal lives, eating fir needles. By delicately nibbling away the green part of the needle, the spine is left as a scrap like the remains of tiny fish skeleton. Voles build nests of twigs, needles, and lichens, some as big as beach umbrellas. They lick the needles to get dew, and can launch themselves from a branch like a flying squirrel. Moving from tree to tree when threatened, they scurry into a hiding place to escape predation with great agility and speed.

When spruce or fir trees are cut naturally, as by beavers, young trees grow back quickly because of the voles. The voles provide a fungal inoculum called a mycorrhizae for the young trees. The fungi live along the roots of the trees and help them absorb nutrients. Clear-cutting and other radical tree removal results in the loss of these microbes. Vole droppings contain the fungi needed by the young plant for healthy growth. So when voles are present, the microbes get reintroduced into the soil and the forest is able to regenerate. When voles are not present, the forest cannot regenerate. Once again, we humans are messing up the cycle. Vole populations are disappearing, just as the spotted owls are, because of the logging of the old trees. Not only are the spotted owls (the predators), the voles (the prey), and the old-growth trees (the habitat) being eliminated, but the way in which the forest recovers is being destroyed. Not only are the creatures of the forests becoming ghost species, but the forest itself is being irreversibly pushed into a state of nonrecovery.

Back in Eugene, we walked along the Willamette River at the end of the day and noticed flocks of English sparrows feeding on the sidewalks. The black-bibbed males and the dull females turned the walkway into a path of undulating gravel. These "germy birds," as described in *Suburban Safari: A Year on the Lawn* (Holmes 2005), are the flip side of rare and endangered. Introduced to North America from Britain in the 1850s, they were thought to eradicate cankerworms and drop worms. It was also believed that immigrants liked to see familiar birds, although this reason for introduction seems a bit far-fetched. The sparrows have become so abundant as to be a nuisance and are now considered invasives. The bully birds have displaced native birds by out-competing them for food and nesting sites, so programs have been established to reduce their numbers. Even birding groups—those folks who love birds with such passion—have suggested ways of discouraging their breeding. Once again, we humans are messing in the business of other species, always tinkering with their lives.

A visit to the William L. Finley National Wildlife Refuge found us again along the Willamette River. The refuge north of Eugene, with its oak and maple woodlands, ash and Douglas fir stands, hedgerow, grass meadows and pastures, contains a small 475-acre undisturbed wet prairie. This tiny parcel is the largest remnant of the Willamette Valley, which once comprised a vast wet prairie ecosystem. Observing the many species of native birds, I thought of this sanctuary as a refuge in the truest sense. It was a place where humans meddled very little in the ways of nature. I wondered if that was a viable approach to conservation: to purchase parcels of lands that are as extensive as possible, make sure they remain undeveloped, and leave things alone. The idea of leaving species alone to make their own homes seemed the right way of doing things.

Of course, conservation always seems to involve managing natural resources, and it is probably naïve to think that we can just leave things alone. We are supposed to think big in terms of eco-system or eco-region management, according to conservationists (Adams 2006). The U.S. National Wildlife Refuge system is an example of protection that clearly has to be managed. Even though a management approach is the way the refuge system is defined, I still like to think that we might be able to set aside as much land as possible in the U.S., leave it alone (as unmanaged as feasible), and let the species thrive without our intervention.

In other parts of the world, land conservation is more compli-cated. According to Dowie (2006), international conservationists attempting to preserve biological diversity by protecting land have displaced indigenous, rural or poor people, and expulsions continue today. While the total area of land under conservation protection worldwide is now estimated at over 12 percent, the cost has been millions of "conservation refugees" who have lost their homes. Some tribal and rural people regard conservationists as just another colo-nial force. Clearly, the issues are not so simple as buying up land and leaving it alone.

Walking along the Willamette our last evening in Eugene, I thought of Kathleen Dean Moore's essays in *Riverwalking: Reflections on Moving Water*, *Holdfast: At Home in the Natural World*, and *The Pine Island Paradox: Making Connection in a Disconnected World* (Moore 1995, 1999, 2004) that speak so poignantly of the vital con-nection to place. She writes of the Willamette as her beloved home. Like the writings of Barry Lopez, who has lived for thirty years along a nearby tributary of the Willamette, her stories remind us that our essential connections have always been with rivers. They always seem like home, in the deepest sense of the word. I won-dered if we might someday think of home not only as our personal home, or our general human home, but in a broader sense, home for all the creatures that share our earth.

Flathead River: Fires in the West

When I looked out the van window and saw the road's shoulder drop thousands of feet into a smoky canyon, I was grateful Ed was our driver. Ed talked about the land with such affection that my queasiness over the heights disappeared. He had grown up around Browning, Montana, was of Irish and Blackfoot descent, and knew the land as home. The Going-to-the-Sun Road in Glacier National Park, one of our nation's most scenic highways, was our last day in the park, and Ed was the perfect driver-guide for the most spectacular ride of the season.

All summer the newspapers reported wildfires in the west, so our trip became one to see this wilderness before it "went up in flames." The Kalispell airport terminal was like a smoking lounge; it reeked of burning leaves. But traveling Highway 2 into the park, the air cleared, the pavement turned a rosy red, and the river sparkled alongside the highway. The North Fork and Middle Fork of the Flathead River form the western boundary of Glacier National Park, and the road to East Glacier followed the middle fork of the Flathead. At several points above the river, we noticed white plumes rising from the mountain ridges, but a southeasterly wind kept the smoke moving away from us.

We arrived at East Glacier in the early afternoon and checked into the Glacier Park Lodge. When I booked the room, I thought the lodge was part of the national park system. (It was linked to the national park service's Web site). I discovered that it was owned by a large corporation called Glacier Park, Inc., headquartered in Phoenix, as were most of the lodges near the park. The company is a subsidiary of Viad Corporation, a $1.6 billion revenue-generating S&P SmallCap 600 company listed on the New York Stock Exchange as VVI. It has a board of directors that includes the president of Johnson Publishing Company, the chairman of Smith International (a company that supplies products and services to the oil industry), a former Vice President of SBC Communications, a former United States Senator, an economist on the Board of Governors of the Federal Reserve, and a long list of moneyed people. What a surprise to learn that all the lodging adjacent to one of our country's most beautiful national parks was corporate owned.

Later in the afternoon, we drove Highway 2 to the town of Browning. Traveling east from East Glacier, the mountains faded in a few miles, and the short-grass prairie stretched out onto a vast plain that ended in sky. The wind shifted and smoke drifted over the sun, turning it into an orange baseball. The smoke-tinted landscape became antique gold. Prairie potholes adorned with dark green needle grass and western wheatgrass teemed with birds, so we pulled off the road to watch yellow-headed blackbirds and chestnut-collared longspurs balance on the reeds. The smoke took the sharp edges off the landscape, and in the dim light everything looked like an old painting with an aged and yellowed patina.

This place where mountains meet prairie is known as the Rocky Mountain Front. It runs about 150 miles from the Canadian border south to around Helena, Montana. Along it lays lies the largest intact wild grassland ecosystem in the United States—about 1.5 million acres; it has been called "America's Serengeti." But the region is under intense pressure from the oil industry; geological evidence suggests there might be oil reserves beneath its lovely grasses. Like the buffalo and the Plains Indians who once lived here, the wilderness of the Front may disappear. That it would be by the hands of the great-great-grandsons and grand-daughters of those who eliminated the buffalo and the Plains Indians seemed if not regressive at least consistent.

The town of Browning, the largest in the Blackfoot reservation, consisted of a few streets lined with tiny houses and dirt yards. Most of the homes were less than a thousand square feet and looked like trailers; in fact, many were trailers. At the local grocery store we discovered a parking lot full of rusty vehicles. Not a new car in the bunch and most were old battered pickups. An elderly Indian man with missing teeth wearing a worn flannel shirt put his box of oatmeal, bottle of milk, and bag of coffee on the counter and counted out his change, a couple pennies short. The girl at the cash register put his groceries in the plastic bag, nodded him on, and began ringing up the purchases of the next person in line. The next customer was a middle-aged, full-figured Indian woman whose worn purse and ill-fitting sweater looked as tired as she did. She kept pushing a strand of loose hair back away from her face as if that small gesture of organization would solve all the problems of her world.

We returned to the lodge, where well-groomed guests meandered about the lobby in their outdoor hiking outfits or sat sipping bottled water while reading the *Wall Street Journal*. The contrast between the local folks who lived in Browning and outsiders who visited the lodge was obvious, and anyone who argued that America is not a society of classes based on wealth would be hard-pressed to make their point having experienced this contrast.

The next morning, we brushed the black soot off the car and headed for the heart of the park, Many Glaciers. The Swiftcurrent Nature Trail begins as a path through a stand of lodgepole pines. Engelman spruce and subalpine fir also grew along the trail, and some sections of the trail had patches of bear grass. Like huge coned dandelions, the flowers were white and full, although some were withered and slightly past their prime. The trail ran along Swiftcurrent Lake, joined the Lake Josephine Trail, then the Grinnell Lake Trail, and finally the Grinnell Glacier Trail. Instead of a quiet walk, the trails were full of day hikers, backpackers, and families with kids—some of them really noisy. Perhaps they were taking the suggestion of the rangers who advised making noise to alert the bears, but it was not the way I wanted to experience the trail. After an hour of stepping off the trail to accommodate one group or another, we turned around and returned to the trailhead.

Along Highway 89, several red buses came and went. The buses, called "Reds," had propane engines, which meant cleaner emissions, and seemed like a sound conservation practice. With a little Internet searching, I discovered there was more to the story. The thirty-two buses, made by Ford Motor Company, had been donated to the Glacier Fund as a tax write-off. They were then filmed for use in a TV commercial that touts the virtues of Ford as a company that is concerned about the environment (another tax break). The Glacier Fund, a nonprofit (tax exempted) organization "dedicated to the preservation of Glacier National Park through private capital fundraising and other partnerships" has a board of directors that includes a long list of the very wealthy, from Louis Hill, grandson of the railroad baron who was president of Great Northern Railroad, to John Teets, CEO of Dial Corporation, which is now reorganized as Viad, the corporation that owns the lodges in the park.

The Glacier Fund is part of a larger organization called the National Park Foundation (NPF), a model of corporate tax-free contributions. The NPF has partnerships with Kodak, American Airlines, Discovery Communications, Time Magazine, and Aventis (the maker of Allegra). The foundation's Web site claims to have brought in $42 million in contributions in 2002 with an administrative cost of 10.6 percent. Its Board of Directors is comprised of moneyed people such as Bruce Benson, president of Benson Mineral Group, Inc; Spencer Eccles, Chairman of a large banking group; and Linda Fisher, former VP of Monsanto, to name a few. One interesting item caught my eye. One of the grants given by the nonprofit foundation went to its "partner" Discovery Communications to make the films titled "Caves" and "Wildfires." In short, the media corporation gives money as a charitable tax write-off, then gets it back as a grant to make nature films which go to support the company. I can't say these people aren't shrewd money handlers.

The next day, we signed up with Sun Tour, the only locally owned company in East Glacier, and traveled the Going-to-the-Sun Road with Ed. The road runs east and west through the park and twists through some of the most spectacular mountain scenery in America. We stopped along St. Mary Lake to see the most recognized image of the park, the one captured by Ansel Adams. The sky was slightly smoky, and the grey patina seemed to put us right in the center of that Ansel Adams photograph.

A brown dot moved across the mountainside far off in the distance, and Ed identified him by the hump on his back as a young male grizzly foraging, eating berries and fattening up for the winter. Ed said grizzly bears once roamed North America from Mexico to Alaska and as far east as Minnesota. Today, outside of Alaska, there are only two significant populations of this endangered species: one of about 400 to 600 bears in Glacier National Park and one of about 500 to 600 bears in Yellowstone National Park. Three other tiny populations occur in the Northwest: along the Idaho-Montana border in the Cabinet-Yaak Mountains, in the Northern Cascades of Washington, and in Selkirk Mountain in the panhandle of Idaho with population surveys reporting only about thirty or fewer bears in each location. Grizzlies are a popular topic among writers, and much has been written on the general biology, conservation, history, and human encounters with these awesome creatures (Busch 2001, Bass 1995, Peacock 1996, Murray 1992). We watched the bear through our binoculars for a while until he finally disappeared into the bushes.

Along a narrow twisting section of road, dense smoke engulfed the road. No flames were visible, but columns of white smoke billowed out from a few burn spots. The fire lines were short and the burning areas were small for the amount of smoke. The burns formed a kind of mosaic, a patchwork quilt with small sections of blackened forest among the green. Ed told us how the ecology of the region was fire-dependent, and how fire was a natural part of the land's dynamics. The seeds of many plants depend on fire to germinate. When cellulose burns, a compound called gavinone is produced that acts like a plant hormone and stimulates seed germination and plant growth. Fire also releases nutrients that promote plant growth, and the new growth after a burn provides good grazing. In short, fire was a natural rejuvenating force, Ed explained.

Some of the best writings on fire ecology are those by Stephen J. Pyne, who has written more than a dozen books on the subject (Pyne 2001, 2004). He explains that fire gets such a bad rap because it destroys homes and buildings. People have built homes, lodges, and businesses out in the wilderness. When a natural fire occurs because the ecology of the forest involves a fire cycle, property owners want it put out. Policies of fire suppression result in more available fuel and thus more intense fires. When a wild fire eventually occurs in an area not burned naturally for a long while, it burns hotter and more ferociously. Fire gets blamed as the great destroyer, but in truth, fire has always been part of the natural biological cycle. It is humans that have interfered with that cycle.

After crossing the Continental Divide, we began the descent down the west side of the mountains. It was wetter and Douglas firs lined the road. In their mossy green raincoats, the trees resembled those of the temperate rain forests of Olympia National Park. At McDonald Lodge, we stopped for lunch and noticed firefighters in bright yellow slickers roaming the parking lot like high school bands getting ready for a parade. Of course, their work was far more serious than a parade. Sometimes these men and women lost their lives fighting fires. Maclean in *Fire and Ashes: On the Front Lines of American Wildfires* (2003) presents an account of the work these courageous men and women carry out.

When we entered the lodge restaurant, the dining room was hazy with smoke and I could not resist asking, "Could we have the nonsmoking section, please?" The waitress must have heard that remark a thousand times, because she just rolled her eyes and gave a half-hearted smile. After lunch we stood on the edge of the lake and tried to imagine the shore. With visibility only a few feet, we could see nothing but grey. It was like being inside a cloud.

The return trip through the park back to East Glacier was as spectacular as the going trip, with different views of the glaciers and mountains. While viewing Jackson Glacier, Ed remarked that Glacier National Park had more than 150 glaciers when the park was first established in 1910, but today only 27 remain. Many glaciers have become ice patches, and those remaining have lost 80 percent of their ice sheet due to global warming. By 2030, he said, park scientists predict there would be no glaciers left.

The next day, we left the park and returned to Kalispell. The river ran along the highway like a joyful song. A bald eagle perched on a branch overlooking the river, and mountain goats hovered like wisps of white clouds on the cliffs above it. Through our binoculars we saw their great shaggy coats hanging down in tatters. We stopped for one last look at the river and to watch the white waters rush over the granite boulders. The air smelled clean, with no hint of smoke, and it was simply glorious along the river banks. Yet despite all the majestic scenery—the river, the sky-scraping mountains, and the steep valleys—I could not shake my gloomier thoughts of how corporations have control over this national park.

To get to Montana, I had to fly on American Airlines whose parent company is one of the biggest corporations in the United States, AMR. The rental car was from National, a subsidiary of a company called ANC that generated revenues of $2.4 billion in 2002. Transportation aside, to be near enough to get into the park, I had to stay in a lodge owned by Glacier Park, Inc. (Viad), a corporation that generated $856 million in annual, 2006, revenues. To visit my country's premier national park required going through large corporations.

The realization that corporations control access to this park and many national parks and public lands was not a comfortable thought. What little wilderness that remains in America being accessible only through corporations did not seem congruent with the concept of public lands. I fear that corporations will control all our wild lands in ways that will not benefit the common citizen. They may be developed, drilled on, clear-cut, and exploited in other ways. The facts are quite clear that those who benefit from corporate profits are few. It is common knowledge that CEOs make over 350 (sometimes 700) times the income of the average worker, and such inequality leads inevitably to exploitation. Perhaps I would not mind quite as much if that old Indian man at the grocery store in Browning had a comfortable home, a dependable truck, warm clothes, and a dental plan as good as Mr. Teets of Viad, Inc., or Mr. Hill of the Glacier Fund. No, actually, I would mind. I do not want my country's public lands controlled by corporations.

The Flathead River's question seemed to be "Whose woods are these?" I fear they may be corporate America's woods, and corporate America has shown a pattern of exploitation of its workers, avarice and dishonesty by its leaders, and disrespect for our country's natural resources. The very glaciers that created the park are melting because of industrial emissions and corporate indifference. Why should we expect corporate America to respect and honor our nation's land? I fear the fires in the West are not just the forest burning. The more deadly fires are those of corporations who would exploit our nation's land to make the few wealthier.

Tensaw River: Longleaf Magnolias

If I believed in ghosts, I might be inclined to think that is what I saw in Blakely State Park. There was morning fog, and we were walking the Breastworks Trail near the old battlefield where the trail dips and runs through a swampy section. I caught a glimpse of something that looked like a human figure. As we walked a little further, I lost track of it and never saw it again. It was probably just an old tree stump.

Blakely State Park is the site of the last battle of the Civil War. It is known as Fort Blakely, although nothing remains of the old fort now. The irony of this battle is that it was fought on

April 9, 1865, six hours after Lee surrendered and the Civil War officially ended. So the 3,500 young men who died there, both Union and Confederate soldiers, died fighting a war that was already over. This fact seems so tragic. That's not to say the whole war wasn't tragic, but this loss seems especially heartbreaking.

We were doing some native plant spotting in the old battle-site-turned-park on the Tensaw River, described and photographed in *In the Realm of Rivers: Alabama's Mobile Tensaw Delta* (Walker and Holt 2004). Blakely State Park is a very old place. It was the site of a prehistoric Indian civilization dating back 4,000 years, the location of an early French trading village, a settlement of New Englanders, and Baldwin County's first county seat in 1814. Like most Native Americans, the Apalachee Indians were displaced and pushed west. A yellow fever epidemic and a land speculation scheme doomed the nineteenth-century town to oblivion as a ghost town.

Today the state park, with its upland live oak groves, river bottomlands of cypress and tupelo, and pine forests, is protected land. Of its 3,800 acres, 420 acres are protected in Forever Wild of Alabama, one of Alabama's most successful conservation programs. First established in 1992, Forever Wild has acquired forty tracts totaling over 100,000 acres, much of it in the Mobile-Tensaw delta. Funded by offshore natural gas lease royalties, enhanced by private gifts, and helped by The Nature Conservancy, Forever Wild has purchased and manages unique lands with the purpose of protecting endangered and rare plants and animals.

We had been coming to the park throughout the year to see the rare bigleaf magnolia (*Magnolia macrophylla*). It was late December, and huge piles of grey leaves were strewn like old rags beneath the bare-limbed tree. The tree, devoid of leaves with its smooth bark and gangly branches, had the look of a Calder sculpture. Without knowing it was a bigleaf magnolia or noticing the enormous fallen leaves, it could easily be overlooked as a common shrub-weed. Unlike its relative, the southern magnolia (*M. grandiflora*), which gets as large as a live oak and stays green all year, the bigleaf magnolia is an understory tree. It sheds its leaves in fall and rarely gets much taller than thirty feet or much wider than a few feet.

In spring we had noticed the leaves emerging from white buds at the tips of the branches. The leaves unfolded like enormous scrolls. Huge, three feet long, they are the largest simple leaf of any plant in the United States. Big floppy things, like banana fronds with ear lobes at the base and fuzzy white hairs underneath, they were shiny green. In a rosette pattern, they looked like giant ceiling fans. The gigantic flowers began to appear in April, and by May the trees along the trail were adorned with porcelain blossoms. The trees transformed the southern woods into an arboreal china cabinet. One gardener's description compared them to his grandmother's big buttermilk-colored soup tureen. Six petals, each about twelve inches long, splayed out in an open rosette. Thick, creamy white with rose-purple blotches in the center, they smelled sweet, way too sweet. The aroma had the sensation of hands clutching my throat in a stranglehold. The flowers bruise easily and turn brown when touched. They attract and are pollinated by rove beetles. These beetles predate bees and butterflies, which did not exist when bigleaf magnolias first grew. Their interdependence with primitive insects is an indication that the tree is a very primitive species. In fact, fossil records indicate that magnolias have been around for more than a hundred million years.

The genus *Magnolia* was one of the first flowering plants to evolve on earth. Of the eighty species native to the United States, many are widely distributed and grown as ornamental plants because of their showy flowers. More than 120 species occur worldwide. While popular as garden plants, some have medicinal uses. The bark of *M. officinalis*, for example, has been used in a Chinese medicine called houpo. Other species of magnolia have been used in traditional medicine for thousands of years, but I could find nothing to indicate whether the bigleaf magnolia had such properties. In truth, I was glad, because collectors would surely strip off the bark and kill the easily broken trees if it was known the plant contained pharmacologicals.

The bigleaf magnolia was discovered in Alabama—in fact, very close to this park—by William Bartram in the 1770s. Bartram wrote in his journal, ". . . how gaily flutter the radiated wings of the Magnolia auriculata, [the nomenclature has changed] each branch supporting an expanded umbrella, superbly crested with a silver plume, fragrant blossom, or crimson studded strobile and fruits" (Harper 1958, 150). A Google search turned up a recurring reference to Andre Michaux, the French botanist, who supposedly first described the plant in North Carolina in 1789.

In September we observed the flowers turning to seed cones. As a primitive plant, magnolias have cones like their ancient companions, the conifers. By October the trees were covered in red fists, and a month later the ground was covered with empty cones. They had been picked clean by towhees, yellow-bellied sapsuckers, red-eyed vireos, and kingbirds. The seeds, having a very high fat content, are a favorite fall food for birds. About the same time, the brown leaves dropped in a display of gargantuan leaf fall. Similar to a sycamore tree that loses its leaves like a giant fish shedding its scales, the magnolia also had a monstrous molt. Now, in the winter month of December, the weathered remains of the fallen leaves were the color of bleached driftwood. In a few months the tree would begin remaking the massive leaves, and the cycle would begin again.

I have often wondered why magnolias are used as symbols of the south. The southern magnolia (M. grandiflora) is the state flower of both Louisiana and Mississippi, although for a time it was not certain whether the magnolia would retain this distinction in Louisiana. A long and bitter battle called "the great state flower debate" raged from 1950 until 1990 over the magnolia versus the iris (Iris giganticaerulea) as the state flower. Members of the Louisiana Iris Society and magnolia lovers engaged in name calling and derogatory remarks until a compromise was finally reached in 1990. The magnolia was designated the state flower, and the iris was given the distinction of the state wildflower (Louisiana State Flower, 2006).

Magnolias also seem to be popular symbols of southern womanhood. When I was growing up, weddings were decorated with boughs of magnolia. They inspired a character in a play we wrote named Magnolia Blossom. She was a silly thing, patterned after the character Prissy in *Gone With The Wind*, except she was white, blonde, and a sorority sister (we were such satirists when we were young). The phrase "Steel Magnolias" has come to mean those seemingly fragile Southern women whose strength of character pulls everyone through hard times. But I have never thought flowers were particularly good metaphors for women, and in truth, I have never known any woman who reminded me of a flower. How certain metaphors get established sometimes mystifies me. But if I were to metaphorically associate anything with the bigleaf magnolia, it would be a place—a place like Blakely State Park.

Bigleaf magnolias are probably the only ghosts that inhabit Blakely State Park, and in a way, they are living ghosts from the Paleozoic past. They come from an ancient time when dinosaurs roamed the earth. They embellish the river lands and give them an exotic beauty. The bigleaf magnolia and the park seemed so well suited to each other. And what a wondrous thing it was that we could watch a year in the life of this lovely tree in its riverside home.

Cumberland River:
Magnolias and Other Rare Plants
of the Appalachians

The trail along the Cumberland River in Cumberland Falls State Park, Kentucky, was leafy with umbrella magnolias (*Magnolia tripetala*) and an occasional bigleaf magnolia (*M. macrophylla*). Many were young trees, no taller than a foot, and it looked as if someone had opened green umbrellas all along the trail. Others were the height of understory trees, about ten to twenty feet tall, and still others reached nearly to the top of the canopy where oaks, beech, and tulip poplar formed dense crowns. The magnolias stretched their slender grey trunks up through the dense canopy, reaching for the open sky with attitude, as if they were entitled to as much sun as the taller trees of the Appalachian forest.

It had rained the night before, and everything was still wet. Silver raindrops lay like drops of liquid mercury on the surface of the waxy leaves. Bright green on top, their undersides glimmered white when the wind blew them over. Where the sun shone through, they had the texture of green lamp shades, and the forest was illuminated with soft green light. In some spots, the leaves cast elongated curved shadows on the open trail. The rain, the heat, and the gargantuan fronds gave the trail a tropical rain forest quality.

The umbrella magnolia differs from the bigleaf magnolia in that its leaves are shorter (about two feet long) and the "earlobe" at the base of the leaf is missing. The six-petal flower is smaller, about six to eight inches in diameter, but like the bigleaf, it is creamy white. The flower smells foul up close, but further away, the odor is more pleasant. The blooms on these trees were too high to be reached and tested, but I wondered how an odor could be foul up close and pleasant from a distance. I learned that chemists have studied the fragrances of many species of magnolias and found they produce mixtures of volatile oils. Analysis of the volatiles of the umbrella magnolia flower shows two benzene compounds called benzenoids—methyl benzoate (55 percent) and acetophenone (36 percent) (Azuma et al 1999). If one of the volatiles traveled farther due to a different evaporation point and smelled pleasant, and the other one traveled a shorter distance and smelled unpleasant, that might explain it. Maybe it had to do with threshold concentrations. A small amount of many compounds, such as some spices, can taste pleasant, while a large amount of the same spice can be bitter or unpleasant. Whatever the chemistry behind the fragrance, I had to remind myself that those flowers were not designed with my nose in mind. They were meant to entice Nitidulidae beetles. To the beetles, they must smell wonderful no matter what the distance.

Magnolia flowers, in addition to producing scents, produce heat and food. The flower is a source of warmth, as well as high-protein pollen that beetles eat. A remarkable thing, this flower! Like a neon sign, the magnolia blossom produces a pigment that the beetles see in the UV range as a gorgeous blue. It attracts with color and scent and then provides a warm abode and food. No wonder those insects have a close relationship with the magnolia. Would not any creature that could have dinner in a beautiful, pleasant-smelling, warm, cozy place keep coming back? Their offspring would keep coming back as well. This is about as good an explanation for co-evolution as any, and these beetles have been around for a long time. They predate the evolution of bees, bats, hummingbirds, and most other aerial pollinators, adding further credence to the idea that magnolias have also been around for a very long time. Magnolias are some of the most ancient of trees, having co-evolved with beetles, traveling the long path of biological time together.

Co-evolution is a rather remarkable process. We learn in our first biology class that natural selection is how species evolve. Evolution is often misinterpreted as "survival of the fittest," or the species that is able to adapt, lasts. This implies that evolution is a zero-sum competition where the organism that endures wins, like a sporting event or an elimination tournament. It is sometimes broadly interpreted in human morality as "every man for himself." Co-evolution suggests something far more inclusive than simply winning or losing. It suggests that the process of evolution involves relationships. To imagine that several organisms accommodate and change with each other over millions of years is to consider that commitment is more than a human construct but is a biological reality. Such accommodations among species appeal to me philosophically and metaphorically because they suggest connections that transcend time.

The magnolias along the riverside were just beginning to bloom—it was early June—and the flowers were hard to see from the trail; they were in the very top of the canopy. It was only when we climbed the ridge and stood looking down from the balcony of the lodge that we could see the forest canopy speckled with white splashes. Against the dark green branches they looked like a choir of Carmelite nuns, their habits flared out in white starchy angles. For a moment I imagined sacred music emanating from the floral choir, ancient hymns written by Hildegard of Bingen that praised a god and honored nature in a cappella song.

The Cumberland River begins where Poor Fork, Martin Fork, and Clover Fork merge in Harlan County and runs almost 700 miles west through southern Kentucky and north central Tennessee. It reaches the Ohio River on the Kentucky-Illinois border, making its watershed quite large. A tributary of the Cumberland, South Fork lies in the Daniel Boone National Forest. Since the forest was close by, we drove over to hike one of the trails in the South Fork National River and Recreation Area.

The trail presented us with an endangered perennial herb, the Cumberland sandwort (*Arenaria cumberlandensis*), growing in small tufts about four inches high in the moist, sandy soil underneath some sandstone outcroppings. Sandworts (*wort* is an old Anglo-Saxon word for plant) are in the genus *Arenaria*, which means "sand-loving," and include more than 300 species. Many of them are evergreen. Since they grow well in low-nutrient soils, such as on rocks and in sand, and have appealing shapes with delicate flowers, they tend to be popular with rock gardeners. This particular species flowers in early July, produces tiny white five-petal flowers, and has wiry stems with long, narrow leaves. It has become endangered because of logging, pot (Indian artifacts) hunters, campers, hikers, and other recreational activities. Radical strip mining for coal, as chronicled in *Lost Mountain: A Year in the Vanishing Wilderness* (Reece 2006), has also impacted many of the mesophytic forest species and resulted in their decline.

It is difficult to explain my fascination with rare and endangered plants, especially this inconspicuous sandwort. As a biologist, I know that species come and go. Environments constantly change, and those organisms that adapt to the changes survive. If an organism cannot adapt, it becomes extinct. That is the nature of evolution. If it didn't happen, we humans wouldn't be here. What concerns me about the species today is that we are altering environments too drastically and too quickly. Scientists theorize that earth has gone through numerous rapid, worldwide, periodic extinctions and that we are currently going through the sixth such species extinction (Thomas et al 2004, Koh et al 2004). The concept that my presence would cause another species to go extinct bothers me philosophically. While I am not particularly interested in harvesting a rare plant strictly for its gene pool, for some future biotech application, or for its potential pharmacological use, I am interested in having a world with as much diversity as possible. I do not know how to accomplish that goal, except to observe the natural environment, write about it, and hope that those who can make a difference will help.

Many of the trails in the Cumberland forests were lined with dead yellow pines. One tree looked as if it had popcorn kernels all over its trunk. This was the work of the southern pine bark beetle. The resin leaks were the points of entry where the beetle had drilled into the bark and the sap had oozed out. The southern pine bark beetle invaded the Daniel Boone National Forest in 1999 and in just five years infested over 100,000 acres of pinelands in the national forest. The beetle eats through the bark, lodges in the living tissue of the cambium layer, and eventually kills the tree. Signs along the trails and in recreational areas warn hikers about falling trees. Once the tree is killed by the beetle, it easily topples. I heard a story about some hikers on a nearby trail who had paused to catch their breath for a moment, only to hear a loud crack and the whoosh of a huge falling pine. Had they not paused when they did, that pine might have crushed them. Efforts to control the beetle have not been very effective, but the introduction of a clerid beetle, a natural predator, has been proposed (our tinkering again). Of course, the introduction of any new creature always involves risks. The results of some introduced species have proven disastrous, and many have become troublesome nuisances in their new environment.

Extensive logging and industrial pollution have changed the Cumberland forests (Little 1995). Indeed, all the forests of the world have changed as humans have deforested the earth in a period of probably less than 600 years (Williams 2002), but the fact that this forest was very different from the one in which Daniel Boone walked seemed a bit disheartening. Few old growth stands remain in the area, but one small hemlock stand was reported in Pine Mountain State Resort Park, so off we went to Pine Mountain, pausing briefly to watch two pileated woodpeckers feeding in the pines in front of the lodge.

The Hemlock Garden Trail was a short loop through a stream-carved sandstone ravine. It was dark, with a low, thick canopy of hemlocks and typical Appalachian deciduous trees—beech, white oak, tulip tree, sweet birch, red maple, and hickory species. The dense understory of rhododendron and mountain laurel, with partridge berry, heartleaf, and mosses as ground growers, gave the trail the feeling of a tunnel. The stand was disappointing because the hemlocks were not especially large and did not look more than fifty years old, so we moved on to another old-growth forest, Blanton Forest.

Blanton Forest was what I had imagined an old-growth forest to be. Supposedly one of the largest old-growth forests in the east, it is what ecologists call a mixed mesophytic forest—a type of temperate deciduous forest found in the central Appalachians that has about thirty canopy tree species. Entering the forest was like walking into a sacred building. A profound aura and silence enveloped the space. The tree trunks were huge columns many feet in diameter, holding up a canopy over a hundred feet high. Mountain laurel and rhododendron grew in the understory, and true mosses and club mosses created a thick carpet. Patches of Fraser's or mountain magnolia (*M. fraseri*) and cucumber magnolia (*M. acuminata*) grew along Watts Creek Trail and Knobby Rock Trail. The cucumber magnolia, with its smaller leaf than the umbrella magnolia (about six to twelve inches long), can reach a height of up to eighty feet, while the mountain magnolia is generally shorter (fifty feet maximum), with leaves about the same size as the umbrella magnolia (twelve to twenty-four inches). Like the bigleaf magnolia, the mountain magnolia has "earlobes."

Beneath an old red oak, clusters of something resembling blanched pine cones rose from the leaf litter. It was squaw root (*Conopholis americana*), a parasitic plant that grows from the roots of oaks. Its pale, inconspicuous flowers could hardly be thought of as flowers. Being in the broomrape family (there are 17 genera and about 150 species in this albino family), these plants have no chlorophyll. They parasitize other plants and live off their host's nutrients. Although some broomrape species cause crop damage, most are generally benign parasites.

Another species of broomrape called beech drops (*Epifagus virginiana*) sprouted nearby from beneath, yes, a beech tree. Joan Maloof in *Teaching the Trees: Lessons from the Forest* (2005) tells the story (among many other lovely ones) of how the beech drops evoked the memory of her first experience with plant scientists and how they uncovered the mysteries of the botanical world. A scaly pinkish straw, it would be several months before the beech drop flowered. When it does, it will produce two types of flowers, a "regular" petaled colorless flower along its slim weedy branches and an underground (cleistogamous) flower that is pollinated by ground beetles. The manner in which beech drops connect to the beech tree is a classic process of recognition and penetration that has been described by plant pathologists. Beech drop seeds recognize a chemical produced by the beech tree roots, and as the seed germinates, the drop develops a radicle that grows toward the tree root. The radicle penetrates the tree's root tissue, a nodule forms, and beech drop roots envelop this nodule. This large bulb-like swelling is where the beech drop anchors and the place from which it saps the tree's nutrients. Beech drops cannot live without the beech tree, so the scraggly plant is considered an obligate parasite. Whether beech drops actually contribute in some way to the survival of the beech tree is unknown. Ecologists may discover that they do. Relationships among close plant associates often involve a dual mechanism rather than a simplistic, one giving and one receiving interaction, and in fact, these mutalistic relationships are far more common than true parasitic ones. So far, the beech drop's contributions have not been identified, so they are considered parasitic.

There is something deeply intriguing about these pale leafless plants that sponge off the grand trees of the Appalachian forest. It is easy to admire the beautiful and stunning tall trees, but it takes a certain eccentric leaning to like these never-do-wells plants that live in the dark corners of the forest and make their living in secretive ways. Their bizarre shapes and inscrutable flowers make them almost fungus-like. But like them I do, for they remind me that in the great diversity of earth's botanical world, there is room for just about anything.

Moving on to the Cumberland Gap National Historical Park, we noticed park rangers and volunteers removing kudzu. Accidentally introduced in dirt hauled up to restore the mountainside, the invasive kudzu would smother this landscape if left to its own devices. All over the South this plant, introduced in the 1930s by the U.S. Department of Agriculture for erosion control (tinkering again), has taken a foothold, or root hold. Capable of growing a foot a day, it has wrapped its green tentacles around the landscape. I have seen roadsides that look like green quilts thrown over them. While it has a certain monotonous beauty, the overbearing plant chokes out native species and has become an expensive nuisance. Conservation efforts to irradicate the plant cost millions of dollars annually. It is the classic example of an invasive species of the southern landscape.

The issue of invasive species has been the focus of conflicting views. According to Daniel Simberloff, professor of environmental studies at the nearby University of Tennessee, invasive species cause more ecological damage than global warming and are the second-leading cause of species extinction (Simberloff and Holle 1999). When an invasive plant takes over, a native plant can lose its habitat to the invader. In *A Plague of Rats and Rubbervines*, Baskin (2002) suggests that the costs of damage from such invasives range from $100 to $200 billion annually. David Quammen (1998) first coined the term "Planet of Weeds" to describe the takeover of invasive plants. He cautioned that certain species could homogenize the planet. If left to spread, they might turn the earth into a proliferation of a few plants, and those plants would be the same all over the world. He cites many examples of invasives, from water hyacinth to melaleuca to musk thistle to Chinese tallow, and the list goes on—plants that have wreaked havoc on the U.S. landscape.

In contrast, *Out of Eden: An Odyssey of Ecological Invasion* (Burdick 2005) reports that after fifty years of research by invasion biologists, the data do not confirm such catastrophic claims. Burdick explains that the situation may not be so dire, citing studies that show little difference between ecosystems rich in native species and ones chock-full of aliens. He claims that nature is more resilient than that, and the so-called green cancer, the term once given to ecological invaders, has not proven to be true. He writes that most invasives actually do no harm. A similar sentiment is expressed in *Aliens in the Backyard: Plant and Animal Imports into America* (Leland 2005).

Articles in popular magazines (Swift 2005) often sensationalize the issues and create the idea that invasive species are like a poison or the plague. To label such plants as evil threats seems more than a little silly. In reality, we Americans are invasives; we did not originate on this continent. Certainly introduced species can outcompete native ones and create imbalances, and invasives have to be removed in certain areas where native plants are susceptible to being eliminated. But invasives are just doing what all life-forms do: making more of themselves, and doing it a little—and sometimes a lot—better than others around them.

We left the kudzu cutters and drove to Pinnacle Overlook for a panoramic view at the Appalachian forests. The green mountains dipped and rolled on until they became indistinguishable from the hazy clouds in the distance. I wondered what the forest would look like 200 years from now. Two centuries of logging have changed the original forest of the Gap, and logging continues to alter the forest today. Coal extraction does, too. Disease, perhaps in part because of the monoculturing of pines, is altering the forest as well. So has the presence of campers, hikers, and recreational users. Kudzu may not be allowed to get a grip on the Gap, but who knows what other invasive will come along and alter the hill forests. Whatever changes occur, I certainly hoped that the umbrella, the bigleaf, the cucumber, and the mountain magnolias that grow so lushly, and are so at home along the rivers and creeks, survive. I'm also rooting for the modest Cumberland sandwort and the broomrapes.

Conhocton River:
A New York State of Mind

From Mossy Bank atop Sharp Hill, the Conhocton River is a thin silver S cutting through the Conhocton Valley, one of the U-shaped glacial valleys that are common in the northern moraine of the Allegheny Plateau. In the southern Finger Lakes region of western New York, the river begins around Springwater, flows down from the hills and winds southward past the villages of Cohocton, Atlanta, Avoca, Wallace, Kanona, and Bath. From Bath it pushes across the southern tier, joins the Tioga River at Painted Post and becomes the Chemung. From there it flows on to the Susquehanna and into the Chesapeake Bay.

One of the best trout fishing streams in New York, the Conhocton is stocked by the state with 10,000 brown trout annually. Twenty miles of public access make it a fisherman's paradise and an idyllic place to enjoy the beauty of a country stream. In *The Trout Pool Paradox: The American Lives of Three Rivers*, Black (2004) uses three rivers in Connecticut as examples to describe how some northeastern rivers became lovely trout streams while others turned into toxic waste canals. When the nation changed from an agrarian to an industrial economy, some rivers were impacted by industries that used them as a source of water or as a waste dump. Since few factories were ever located on the Conhocton River, little restoration was ever needed, and the river today is clean and healthy.

The Conhocton flows about forty miles through Allegheny Mountain forests of broadleaf and hemlocks. On some hilltops, open areas called oak savannahs offer contrast to the forest. Four thousand years ago, a drought caused a die-off of forest, and short prairie grasses took over. Early settlers took advantage of these oak openings and converted them to farmland. In fact, much of the northeastern forestland was converted to farmland in the 1800s. The history of New York land use goes like this: prior to the 1700s, the land was predominantly forest and 95 percent was still forest in the late 1700s. By the 1880s, over 80 percent of the land had been cleared for farmland and sheep pasture. At the turn of the twentieth century, only 20 percent of New York remained forest, but the landscape slowly changed again and farmlands reverted back to forests. Today, 65 percent of New York land is forest.

Arch Merrill's *Southern Tier* (1986) gives a history of the region's settlement. It was once home to the Seneca Indians from about 1400 to the 1770s, but the Senecas were driven off by the Sullivan-Clinton slash and burn campaign of 1779. In reprisal against the Iroquois Confederation, which had sided with the British in the American Revolution, Washington issued a scorched earth policy, and the Indian settlements were destroyed. Its true purpose may have been confiscating the land and extinguishing Indian land rights rather than retribution, since Sullivan's men were given land grants for their service in the war, and various treaties forced the Indians to abandon claims to their native lands.

The history of land ownership in western New York involves a number of land companies and transactions. Forced treaties with the Indians resulted in their loss of the lands to British speculators and European settlers. Probably the clearest records began with the purchase of the Genesee tract, which included all of Ontario, Steuben, and Yates Counties and some of the surrounding counties, in 1788 by Phelps and Gorham from the Commonwealth of Massachusetts. They in turn sold it to Robert Morris, one of the wealthiest men in the colonies, who helped to finance the American Revolution. He then sold it to the land company Pulteney Associates, British land speculators who then sold it to settlers, mostly of European descent. Today the state's thirty million acres are about 25 percent farmland, and about 20 percent (or six million acres) in the Adirondack Park (half of which is in Forever Wild). An additional one million acres has been or will be acquired and preserved in wilderness or open-space initiatives. The state's record of land conservation and preservation has been remarkably progressive even if logging and drilling for natural gas still continues in state forests.

By the late 1790s, settlers had begun establishing trading posts, cutting forests for cropland, and raising sheep in the region. The types of crops varied over the next 200 years of farming, but today the chief produce is potatoes, specifically potatoes sold for making potato chips. Two books, *The Potato: How the Humble Spud Rescued the Western World* (Zuckermann 1998) and *Potato City: Nature, History, and Community in the Age of Sprawl* (Leaf 2004), describe the natural history of the potato and how potato-growing communities evolved. Their descriptions could well apply to this valley.

Highway 415 from Bath to Cohocton follows the river and the two entwine like a braid. Near Avoca, where they cross, Hubbard's Farm sits at the juncture. Like so many farms in the area, Hubbard's is a small family farm in "the mucklands" that grows and sells potatoes and other seasonal crops on a small scale. The signs on the market stall notify the passerby of a progression of seasonal crops, from early summer peas and squash to late summer red potatoes and beets and fall pumpkins and gourds. Hubbard's and the other farms of the area could be replicas of Jane Brox's farm in the Merrimack Valley of Massachusetts. Through personal narrative, Brox (2004, 1999, 1995) presents the American farm as a family endeavor that has become marginalized, made almost extinct because of economic pressures, and transformed into a romanticized place of agro-entertainment. Her narrative presents both the factual history of the last 150 years of American farm life and the spirit of these farms and the working communities that grew around them.

Another perspective of the region is contained in a memoir by Morrow (2004). Her memories of growing up in Geneva, New York, are a collage combining the personal and the historical. Unlike Jane Brox, whose parents were immigrant farmers and whose livelihood depended on their farm, Susan Brind Morrow, a linguist, had a more privileged life. Her father was a prominent lawyer and judge in Geneva; her grandfather was the chief counsel and deputy commissioner of education for New York State; and her husband was senior writer-editor of *Time* magazine. Although Geneva is one of the largest communities in the area, it has remained a small town with a population of 14,000. Morrow's writings reveal a home near Hobart and William Smith Colleges and Cornell University's Agricultural Station, a cottage on Seneca Lake, and outdoor experiences such as skeet shooting, boating, and fishing on the lake.

For many years, I was a summer visitor, although I now live in Bath. One of my most vivid memories was picking peas in Hubbard's fields. The fields were wet from dew, and when the sun warmed them the sweet scent of pea vines filled the air. The swollen pods hung tightly to the tangled vines and required some dexterity to pull off, but we soon became proficient at selecting the plumpest ones. With mud caked like chocolate cement on our shoes and the twinge of well-worked backs, we hauled our baskets back to the shed and paid for our harvest. Sitting on the porch, the afternoon passed pleasantly in the rhythm of shelling. With green stained fingernails we prepared the peas for supper. Floating in milk with a bit of molten butter on top, our bowls of peas held beautiful jade pearls. What joy was the sweet green taste of summer! Some of the peas we froze, and on a cold winter day, they would bring back remembrances of the warm fields.

Downtown Bath has two cafes that serve daily plate-lunch specials, and around the corner the bakery is known for its salt-rising bread on Tuesdays and Thursdays. Of course, there is the proverbial Chinese restaurant run by a local Chinese family and the convenience store managed by a local Indian family. A hardware store that has everything under the sun and a post office also make up part of the downtown. The rest of the stores are mostly taverns and lawyers' offices.

The fairgrounds stay quiet most of the year, but in mid-August the Steuben County Fair transforms it into a beehive of activity. The fair has the distinction of being the oldest county fair in the United States, dating back to 1795. With pig races and horse shows, carnival rides and 4-H exhibits, it has remained an old-fashioned county fair, although in the last few years it seems to have acquired a shabby bit of nostalgia. Some of the exhibits appear to be as wilted as the dahlia arrangements and as worn as the frayed edges of kids' drawings posted on exhibit boards. Last year, I noticed one child with a computer that generated a graph of his heifer's weight gain over time, an image that seemed as juxtaposed with tradition as any in today's high-tech world. Still for a couple of weeks in August, the fairgrounds are an old-fashioned celebration of summer.

The Dormann Library, hidden under its canopy of old oaks, is a busy place all year round. Pulteney Square offers benches, a gazebo for evening band concerts, and a farmers' market. Throughout summer and into fall on Wednesday and Saturday mornings, Amish families and other local farmers bring in their produce, and one side of the park becomes a market. On either side of town, two ice cream stands remain quiet early in the day, but as the summer day heats up, folks flock in for soft ice cream, hot fudge sundaes, maple walnut hard ice cream, or one of the many other choices.

At the end of Robie Street in Murphy's cornfield an old barn has stood for fifty years. It is a favorite perch for turkey vultures. Migrating from South America every spring, they make their summer home in Pleasant Valley. The vultures roost on the tattered roof like a queue along the ridge of the old barn. Spreading their feathers to air-dry, their black wings create a decorative wrought-iron railing, a chain of thunderbirds. During the day they glide in the valley's sky, ever on the lookout for road kill and field pickings. I have grown quite fond of those birds and consider them my beloved neighbors.

In October, the town of Cohocton celebrates its annual fall foliage festival. There is nowhere on earth that smells quite like a fall festival in western New York. Foods like salt potatoes, polish sausage, beef-on-a-wick, sugar waffles, and fried dough mingle to produce an olfactory cornucopia. The tree-sitting contest involves a dozen or so hardy folks who sit in the arms of the old maple trees that line the streets. The sitter who stays the longest over the three-day weekend with the least amount of gear and the least amount of time for bathroom breaks is declared the winner.

I have noticed a change in the festival in the last few years. The types of items offered for sale at the festival were once practical items: tools, household gadgets, homemade baked goods, fruits, and vegetables. Today they are mostly T-shirts, jewelry, stuffed toys, and country crafts. The number of booths has also increased, so now it feels crowded and claustrophobic to move through them. And the look of the festival goers has changed. At one time trim, sneaker-wearing, denim shirt and blue-jean clad fellows and gals attended. Today the style seems to be black leather, untidy and obese. At the last festival, a hefty gal with stringy hair and a cigarette hanging from her lips was selling T-shirts adorned with motorcycles and

NASCAR emblems. She hollered at us that the shirts were two for twenty-five. The wholesome look has left, and I wonder if that is not the way of so many public events in American life.

When the leaves begin to turn, we take to the seasonal roads as often as possible, trying to squeeze in every minute of hiking before winter. The red maples, yellow beech, and orange oaks turn the back roads into painted archways. The creatures of the woods and fields begin to move about in preparation for winter. Woolly bear caterpillars that have munched all summer on nettle, plantain, and dandelions begin to wiggle across the roads. Fuzzy palindromes, they cross the roads like red and black squiggles. Over-wintering under stones and leaf litter, they will emerge in the spring as Isabella tiger moths.

The birds become restless. Flocks of Canada geese honk like congested traffic as they move back and forth from Lake Salubria to Weaver's cornfields. An occasional flock of black wing-tipped snow geese glides over. In the hemlocks, white-breasted nuthatches conversing in nasal nays cling upside-down and prod for insects with frenetic eagerness. Eastern phoebes pump their tails, balancing like high-wire acrobats, then flitter out for an insect and return to the same branch to resume pumping again. It is a rhythmic pattern of pump, fly, return, pump. Kingbirds and goldfinch, chickadees and sparrows, all flutter about in search of food. In the open fields, red-tailed hawks soar in loopy circles seeking mice or perhaps practicing for the great spiral south. Everyone needs to fatten up for the long trip south or a cold winter stay.

The Conhocton River Valley, with its small towns and beautiful landscape, has always been a communal and connective place. But in the last few years disruptive forces have been tearing communities apart. Larry Newhart, a local landowner in Hartsville, described at a Steuben County Sierra Club meeting how wind-energy developers have been turning neighbors into bitter enemies. Instead of community involvement and locally owned or controlled wind generating facilities that would serve the local energy needs, provide equitable profits to the community, or at least generate taxes, big corporations have moved in to acquire land for wind farms using divisive tactics.

A dozen or more companies have appeared on the scene, but those of Clipper Windpower, Inc., Global Winds Harvest, Inc., and EcoGen, LLC, have approached certain landowners in the townships of Hartsville, Hornby, and Prattsburgh, where there are no zoning laws. They have offered lease deals of $3,000 to $4,500 per year for each wind turbine erected. Prospective landowners see money-making opportunities in leasing their land to developers. Other landowners and community members, however, see wind farms as a source of noise pollution, degradation of scenic land, and enormous loss of property value. (The grass roots organization Conhocton Wind Watch gives updates on its Web site, http://cohoctonwindwatch.org.) The effect of wind turbines on birds and bats is also devastating (Nijhuis 2006). The corporations have no interest in strengthening local community ties and have pitted one individual against another. The tension has been reflected in a flood of angry letters for and against wind turbines in the local newspapers, and town meetings throughout the region have become bitter battlegrounds.

The corporations are not local businesses. Global Winds, for example, is owned by several former executives of Nordex, a Danish wind turbine manufacturer; EcoGen is a company out of the United Kingdom; and Clipper Windpower, Inc., is a California-based manufacturer of turbines, several of whose executives formerly worked at Enron. Canandaigua Power Partners, LLC, is part of Calpine, a huge energy company whose chairman of the board also serves as director of AT&T, Citigroup, and Halliburton. General Electric keeps coming up as a controlling company, but I was unable to find any documentation of their association. Clearly, these companies have big corporate links.

A conservative e-zine article (Hall 2006) refers to these developers as "Wall Street vultures." Their interests are profits, and profits are enormous because of government subsidies: ". . . wind power developers are skimming millions via subsidies, state mandated quotas, and 'green power' scams" (Komanoff 2006). Subsidies include tax-exempt profits for fifteen years and huge start-up grants. Global Winds, for example, was awarded $4.5 million in 2002 by the New York State Energy Research and Development Authority to develop wind farms. Clipper Windpower, Inc., was given $11 million in grants by the U. S. Department of Energy and the California Energy Commission for developing wind turbine technology. Other subsidies include a 1.8 cent tax break for every kilowatt hour of electricity generated and sold to the grid. The current rate of electricity is about 3 cents per kilowatt hour, and a 5MW turbine can generate about fifteen million kilowatt hours per year. That translates into nearly half a million dollars a year. Depreciation exemptions result in minimal actual maintenance costs for six years or longer. Once depreciation costs and various tax breaks expire, the corporations will likely sell off the wind farms with no responsibility for the maintenance or presence of the enormous 400-foot metal turbines covering the hillsides. These corporations intend to make huge profits at taxpayers' expense with no responsibility to the community where they farm the wind.

At a time when alternative energy sources are needed, such events are a sad commentary on the political and economic forces at play. When an alternative energy source such as wind could offer a partial solution to the nonrenewable energy crisis, greedy corporations seem determined to make huge profits at taxpayers' expense by developing wind farms in a way that disrupts local communities. Not only do they seem determined to turn some of the most beautiful agricultural countryside in the United States into fields of steel towers, but more damaging, they are turning neighbors into bitter enemies.

The state of New York begins with one of the most inclusive cities in the world, New York City, and fans out into what is referred to as upstate New York, with its small towns including those along the Conhocton River. New York has accommodated every immigrant group and every culture on earth. Every type of community and ethnic group in America can be found in this state, from Chinese and Indian immigrants to the Amish, from the disappearing farmer to the wealthy elite, from the Southern visitor to the modest local landowner. I think the phrase "a New York state of mind" implies that inclusiveness and diversity. Whether you need to belong to a small town or a metropolis, the key is that everyone should have somewhere to belong. Corporations that enter and disrupt our communities and our essential ways of connecting destroy our essential sense of belonging. If we lose that connection, we lose what makes us great as a state and a nation.

Penobscot River:
In Search of Moose

The west branch of the Penobscot River runs along the southwest border of Baxter State Park. We had traveled to the park to see moose. Arriving at the mill town of Millinocket, Maine, on a cold, overcast July afternoon, we checked into a local hotel and found a place for supper. The Appalachian Trail Café had all the familiar comforts of a New England diner: great sandwiches, the nicest waitresses, and the homey background of chatting friends. The next morning we discovered their pancakes were about the best in the world as well.

The eighteen-mile paved road from Millinocket to Baxter State Park became a dirt and gravel road at the Togue Pond Gatehouse, the southern entrance to the park. In keeping with a policy of leaving everything as natural as possible, there are no paved roads in the park. When Percival P. Baxter, a former Maine governor, donated the land to the state in the 1930s, it was agreed that the park would have no electricity, running water, or other modern amenities. A single dirt perimeter road with a few side roads to a couple of campgrounds encircles the park. The only other road runs north to the Roaring Brook Campground near the base of Mt. Katahdin. Of the 100,000 visitors each year, about half climb Mt. Katahdin, the northern terminus of the Appalachian Trail.

The park is part of a larger region in north-central Maine once covered in glacial ice. About 12,000 years ago, it began to melt, and tundra evolved into boreal forest. The Great North Woods of Maine today fall into three distinct forest types: red spruce-balsam fir, northern mixed hardwoods/hemlock (yellow and paper birch, sugar maple, beech, black ash), and white pine. For nearly 300 years (since the early European settlement), the woods have been logged, and little pristine forest remains. A few old-growth forests of red spruce and white pine can be found in the Debsconeag Lakes region on the southeast border of the park, but the Great North Woods are second-, third-, and fourth-generation forests.

In 2002, The Nature Conservancy negotiated a controversial deal with the Great Northern Paper Company (Woodard 2004). Loaning money to keep the company from going bankrupt, The Nature Conservancy obtained 41,000 acres in the Debsconeag Lakes region for a preserve and obtained easement rights to an additional 200,000 acres. The easement agreement restricted commercial and residential development of these 200,000 acres but allowed logging to continue. With that action, more than 750,000 acres of Maine's woods were connected and, in part, conserved. In 2004, The Nature Conservancy also negotiated a plan with Plum Creek (a timber company-turned-real estate investment trust) to protect another 413,000 acres of Maine forestland (Woodard 2006).

It was argued that the reason for protecting such large tracts of land was to maintain forest diversity; larger land areas had to be contiguous. Adams (2006) describes this need for continuous connectivity in *The Future of the Wild: Radical Conservation for a Crowded World*. But the old controversies remain, even when given new names like "easement," and the issues of sustainability still create tension among environmentalists.

The Nature Conservancy claims that the issues come down to how to protect the forest's integrity while harvesting its trees; it is assumed that harvesting trees is the only way to save and sustain the local economy and that logging is better than real estate development. The Nature Conservancy's solutions—easements, compromise, and "partnerships" with timber industries which are now selling off their land to developers—may be the most expedient way of protecting the northern forests, but it is a compromise. In my ideal world, forests would be protected by eliminating massive timber harvesting, using only selected cutting and limited harvesting, and never used for building subdivisions. But in the real world, when profits are not high enough, timber companies just move elsewhere or sell the land to developers. Some timber companies have moved to Canada, South America, and third-world countries where land is cheaper and easier to exploit. Timberlands are now more valuable as real estate, so the lands are being sold to investment entities. While the motto, "Be Less Bad," as Adams points out, is not an especially good policy, The Nature Conservancy's compromise may be the most practical immediate way to protect the northern forest lands from development.

In the morning light, Togue Pond reflected a blue sky and clouds like scoops of vanilla ice cream. We stopped to watch a moose feeding at the water's edge. In our binoculars we could tell it was a cow. A fellow who walked up beside us as we observed the moose, nearly scaring us silly, he was so quiet, said he had seen a calf. We searched for a while but were never able to glimpse the calf. The cow, however, was quite visible and grazed like a suction pump. A cow can eat hundreds of pounds of aquatic plants a day (2.5 to 3.5 percent of her weight) when nursing a calf, and it seemed she was intent on consuming her daily quota of greens.

The section of the Appalachian Trail we walked was in the shadow of Mt. Katahdin. The trail ran along clear streams, ponds, and bogs, crossed bridges of weathered wooden planks, and joined an occasional gravel access road. The bogs trembled with butterflies, dragonflies, and damselflies. Black-winged iridescent dragonflies the size of darts zoomed among the grasses and white arrow leaf. Resting briefly on the tips of seed heads, their aquamarine bodies swayed in the slight breeze and cast shadow puppets on the surface of the water. Black and white swallowtails hovered on water lilies, and yellow swallowtails shimmied up to milkweed. They would flutter off nearly as soon as they landed. Mottled grey-brown butterflies also flitted about, and we wondered if they were the endangered Katahdin arctic butterfly (*Oeneis polixenes subspecies katahdin*). Not being knowledgeable lepidopterists and lacking a field guide, we had to be content to just wonder. Mushrooms sprouted from the spongy soil like white saucers, miniature orange water towers, and white-capped pipes. Sensitive, royal, and cinnamon ferns poked up to add their foliage to the ground décor, and mosses grew with hometown confidence nearly everywhere.

Henry David Thoreau, in *The Maine Woods*, wrote of his three expeditions into the region in 1846, 1853, and 1857: "The primitive wood is always and everywhere damp and mossy, so that I traveled constantly with the impression that I was in a swamp . . ." (Thoreau 1983). Indeed, our trail did run through swamp, but the blue hump of Mt. Katahdin was a reminder that this was not the flatland swamps of deltas and deposition. This was a northern inland swamp with a mixture of lowlands and highlands. The blue mountain, rising 5,200 feet into the clouds, formed a high ridge on the horizon with the other encircling peaks. The contrast between

lowland bog and mountain ridges gave the wetlands a very different texture than a southern coastal swamp.

Along the trail where pines grew, a ghoulish plant rose from
the litter. Pink pinesap (*Monotropa hypopithys*) emerged like a
mushroom, except it was not a fungus but a parasitic plant. A few
inches tall, it resembles its close relative Indian pipe (*M. uniflora*),
but instead of a single flower, a cluster of flowers grow at the end
of its bent stem. Its waxy pink tint made me think of little girls'
painted fingernails. With its constant companions, white pine (*Pinus
strobus*) and a mycorrhizial or root fungus (*Tricholoma* species), the
three make for an interesting ménage a trios. Studies by ecologists
(Bidartondo and Bruns 2001, 2002, Bidartondo et al. 2001) have
shown a remarkable and long-term (co-evolutionary) association
between these three very different botanicals. The chemistry of their
interdependency is not completely known, but radioisotope experi-
ments have shown that carbon flows from tree to fungi to pinesap
in an intermingling of nutrients. The pinesap gets nutrients from
the fungus and the fungus may benefit from the pinesap in some
way, thus making their relationship a mutualistic one, so the plant's
description as a parasitic plant may not be accurate. The relationship between the fungus and the pine tree has been known for
some time. Each provides nutrients to the other in a classic mutualistic fashion, the mycorrhizal fungus also offers the pine some
protection from pathogenic fungi. Furthermore, the threesome is
very specific: pinesap only lives with that particular species of fungus, and that fungus only associates with that species of pine. There
may be others involved in the community, such as helper bacteria,
but their micro-level involvement is more difficult to demonstrate.

As I looked at the tiny pink flowering pinesap and the nearby
tall pines, I imagined a massive fungal wiring below ground connecting these two "Mutt and Jeff" plants, a wiring that indeed
connected all the plants of the forest. How amazing the interconnections are between such different plants in the north woods. Yet

when the forest is clear-cut, one of the first species to disappear is the pinesap (Roberts and Zhu 2002). Even when the forest is replanted, the pinesap never returns. I wondered, with our desire for toilet paper, disposable paper plates, and all the other myriad wood products we seem to need, how we lose sight of the demise of the inconspicuous pinesap. How is it that our presence seems to always leave the legacy of loss?

Driving to Kidney Pond, we encountered a carload of campers who said a bull moose was feeding in a nearby pond. We drove slowly and carefully, looking in all directions, and even though we knew he would be there, it was still startling to see such a colossal creature. His enormous head, blurred by a cloud of insects, was adorned with a massive rack. When he lifted his head, he seemed to pull the surface of wet earth up with him like a blanket. Dripping with pond weeds, head like an uprooted tree trunk, he was a behemoth, a creature of the imagination come to life. He was the bog Minotaur.

We stared at him for as long as we felt we could without becoming a nuisance and finally moved on. At the end of the road, a campground with a few cabins lined the lake. Checkerboard loons bobbed like corks in the far corner, and on the back of one, two babies rode a parental flotilla. Another moose and her calf were feeding along the shore, but we were never able to get a good look at them. They kept moving away into the thickets. As we walked a path that followed the lake, trying to get a glimpse of the creatures, we noticed a little girl and her mother. The little girl had been swimming in the lake and her mother was bent over her with a rinse bottle, washing her feet. They were preparing to go inside for the evening, to fix dinner, and to settle into night. In the background, we could hear the heavy slush of the moose and calf moving through the muck at the lake's edge. How poignant the sights and sounds of nativity are in the dusk of the northern forest.

Returning to our lodging in town, we bemoaned the fact that there were so few cabins in the park. Reservations for them fill up the first of the year, and people wait for years to get a cabin. For the next several days, we traveled in and out of the park, and every day we discovered something new and fascinating. As our time came to an end in the great north woods of Maine and the Penobscot River, I thought how pleasant this kind of biology is. This was old-fashioned biology, the philosophical kind that gave meaning. Of all the disciplines in biology, being a naturalist and getting out into nature was the best, even if it is dismissed as nonscience and not even considered biology by many. Molecular biologists are certainly the most prestigious, on the top of the status pyramid, and they have been very clever in answering relevant questions. They have reconstructed the biological past with sequencing techniques and built phylogenetic trees that show chemolithotrophs as the life-starters. They have constructed theories that describe the merging and associating of bacteria and cyanobacteria, creating fungi, algae, plants, and animals, and when that mixing and meshing occurred, life forms exploded and covered the earth. With their nucleotide sequencers, they have worked out what came when, what the biological continuum probably was, and how life evolved. Our most eminent modern American biologists—men such as James Watson and Craig Venter of the Human Genome Project—have reduced life to the defining units of genes and DNA, yet I wonder why their brilliance offers so little in the way of hope or comfort. Maybe I just have trouble with biological reductionism giving unifying answers, and with who gets to narrate.

Field biologists and ecologists have described fairly well what is here now, and much of our present-day biology continues to be revealed. We have surmised much of our past and much of the present, although I do not want to imply that we know it all. The tough part is the future. Our scientists have not been able to tell us much about that, except that if we don't stop polluting the earth, cutting down the forests, and destroying our natural habitats, we will probably eliminate too many life-forms, ourselves included.

In a time when many of my friends and family are battling cancer or other serious illnesses, where their lives and mine are approaching that great wall of old age and death, and when our personal futures seem uncertain, I think about that. Sometimes that future looks imminently like an end. Some of my friends find hope in the belief that a personal god looks out for them; some find comfort that god exists at all. Some find comfort and hope in the belief of an afterlife. Some find comfort in the here-and-now and in the natural world. The comfort I am able to muster involves knowing that there are places where wild creatures live and that their lives go on, that there are places where life is most beautifully entangled in the complex natural relationships of plants, in the mother and child connections, in all the great diversity and wildness of natural habitats. Hope, for me, is a thing with feathers and fur and bark and cell walls; it is all that lives and will continue to live.

Klamath River:
Forest Monoliths

The Klamath River winds through 200 miles of coastal mountains in Oregon and California before it empties into the Pacific Ocean. It is one of the country's most beautiful rivers and part of the National Wild and Scenic Rivers system created in 1968 by the Wild and Scenic Rivers Act to counteract the damming, dredging, and degrading of American rivers. The Klamath's watershed has been well chronicled (Blake and Blake 2000, Wallace 2003, Weidensaul 2005). A Google search of "Klamath River conservation" gives 440,000 sites, which demonstrates the enormous amount of literature available on this river. Of all the excellent writings on the Klamath, the words of Louise Wagenknecht seem the most poignant. She writes, "I was born into the generation for whom the roads were built, the rivers dammed, the old growth cut, entire races of salmon destroyed, and ten-thousand years of wealth consumed in a moment of time" (Wagenknecht 1998, 97). When I consider that it is my generation she refers to and that my peers are most responsible for the destruction of the rivers, it shocks me. I had never really considered the enormous environmental destruction caused by my friends, my family, those who share my time on earth. Somehow, her words made it all seem personal.

Where the Klamath empties into the Pacific Ocean about twenty miles south of Crescent City, California, a group of national and state parks cluster around its terminus. The parks are part of a string of preserves, recreational areas, wilderness areas, refuges, and sanctuaries that make up a 450-mile fog belt known as the Redwood Coast. Where the Klamath River ends, the redwood forests grow.

My traveling companion and I were in Redwood National Park. Having driven up the windy Bald Hills Road, we planned to walk the nature trail in Lady Bird Johnson Grove. It was foggy and shafts of sunlight streamed down through the columned trunks. The forest was embraced in fog, which seemed appropriate for our first images of the redwoods because these giants depend heavily on fog. Described in Noss (1999) and Barbour et al (2001), coastal redwoods (*Sequoia sempervirens*) are able to create their own rain by condensing fog on their needles and funneling it down to the ground as rain. Not only are these trees the tallest on earth—reaching heights of 350 feet—and not only do they live very long lives—some as old as 2,000 years—and not only do they produce the most biomass of any plant on earth—in excess of 85,000 tons per acre—but they use the most ephemeral form of water; they use fog. In fact, 30 to 40 percent of the trees' water comes from fog.

The idea of converting fog into tree mass seemed on that morning an absolute miracle! As a biologist, I know that plants use sunlight to convert water and minerals into plant material. That had always been an obvious, but rather abstract fact, a set of flow diagrams on a chalkboard or a black and white equation in a textbook, until that morning. Walking among those titans in the fog gave a whole new meaning to the phrase *aquatica metamorphosis*. The fact that fog gets turned into liquid water, that water gets pumped up enormous heights into the tree crown where sunlight converts it into sugars, and that those nutrients are pumped down and around to make trees the size of water towers seemed a magnitude beyond miraculous. In a lifetime I have grown an average body; I will be lucky to create a few books (with the help of trees), and that's about all. But a single redwood tree creates in its lifetime a living structure of gargantuan proportions. The reality of that much creation almost overwhelmed me.

The plaque at the trailhead read, "Dedicated to Lady Bird Johnson by Richard Nixon, 1969." While I never imagined either one of these public figures as notable environmentalists, the record suggests that they did play a role in conservation. Gould (1988) in *Lady Bird Johnson: Our Environmental First Lady* describes the first lady's many efforts at preservation. I had never given much credence to her as a progressive conservationist and thought of her more as the lady who planted wildflowers along the roadside. I'm afraid the impersonation of Lady Bird as the superficial Southern gardener in Fannie Flaggs' stand-up comedy routine made a lasting and unfair impression on me. Lady Bird was more than the highway beautification queen. Her efforts in founding The National Wildflower Research Center were worthy of a true conservationist. The Wildflower Center's projects involve plant databases, programs to protect endangered species, seed banking, and habitat restoration, all of which are important missions in modern conservation.

The story of preserving the redwoods did not end with Lady Bird's work in the 1960s (Schrepfer 1983). It continued well into the 1990s with incidents such as Redwood Summer. The events of that summer have been described in *The Last Stand: The War Between*

Wall Street and Main Street Over California's Ancient Redwood (Harris 1997) and *The War Against the Greens: The Wise Use Movement, the New Right, and the Anti-Environmental Violence* (Helvarg 1997). Environmentalists from Earth First! and several other organizations tried to stop the cutting of the ancient redwood trees by big timber companies. Georgia Pacific, Louisiana Pacific, and Pacific Lumber had been taken over by corporate raiders, and their actions polarized the forest conservation issues, raising tensions between environmentalists and loggers/timber companies. Propaganda by a so-called grass-roots organization called the Wise Use Movement created even more tension, and protests by environmentalists, such as chaining themselves to logging equipment, sitting in trees, and blocking logging roads, were met with violence and the arrests of many activists. The most violent incident was the car bombing of Judi Bari, one of the primary organizers of Earth First! She was badly maimed, nearly killed, and then further victimized by smear campaigns and accusations that she had planted the bomb herself. Even after her death from breast cancer in 1997, and after a federal jury in 2002 returned a verdict in her favor in a lawsuit against the FBI and Oakland police officers, Judi Bari continues to be the target of right-wing anti-environmentalist groups. *The Secret Wars of Judi Bari: A Car Bomb, the Fight for the Redwoods and the End of Earth First!* (Coleman 2005) exemplifies the continued attempt to make Bari look villainous by implicating that she and other environmentalists were dopers or communists or crazies. Published by the conservative Encounter Books, the book is full of inaccuracies and sensationalism, according to several reviews. Encounter Books, with its history of smears on Hillary Clinton, Al Gore, and Noam Chomsky, seems determined to continue the fight over who controls the trees. But their battle is not just over trees; it is a struggle for control of information. Those who control our stories control not only the trees but all the earth's creatures.

The well-maintained trail through the grove was a quiet pathway, seemingly far removed from the old conflicts. Its avenue of living columns and open understory allowed us to see the ancient trees in all their glory. They were stunning trees, and it was a stunning forest. Looking at the thick, fire-scarred trunks and then up into the massive branches of the canopy, containing an entire ariel epiphytic ecosystem of huckleberry bushes, lichens, rare mosses, and ferns (Preston 2006, 2007), I thought how right biologists were to use the tree to represent all life. "The tree of life" metaphor is used to represent the evolution of life and the interconnectedness of all creatures, beginning with the primitive archebacteria. The long redwood tree trunk seemed ideally suited to signify the life and times of these primitive microbes, for they were around millions of years before higher forms appeared. From the long trunk of microbes, the limb of the algae branches out to form the plant kingdom, and the pattern expands. Another limb, indeed another trunk, opens to form the protozoa and then the fishes, reptiles, birds, and mammals of the animal kingdom. Limbs branch out into more limbs that branch out to even more limbs. It wasn't a random metaphor when biologists decided to use the tree to represent the dynamic process of life's evolution. Although we know that mergers were probably the driving force in the evolutionary process, what better symbol for understanding the connection between all living organisms than the tree? The very essence of a tree's structure reflects the way organisms are connected, and like all life, the process is ever expanding, ever branching out, ever broadening, and always moving in new directions.

There is probably nothing more life-affirming than a tree or a river. In *Tree: A Life Story*, Suzuki and Grady (2004) write that trees are like rivers with their patterns of branching and expansion. Where the Klamath River ends and meets the sea, the tallest trees on earth rise from the mist. Perhaps the grand redwoods serve to remind us of our humble beginnings and our present status as only a tiny branch on the tree of life. Perhaps the river and the trees tell us to be more mindful of our moment of time. Perhaps they serve to remind us that if we, specifically my generation, are the ones who have wreaked such havoc on the earth, then we are the ones who must restore it. The storyline of the Klamath River, a river that ends in trees, may be that we must do what we can to ensure that all life (not just our own species) goes on and on.

Ellijay River: Applesauce

In northern Georgia, about sixty miles north of Atlanta in the foothills of the Appalachians, the Ellijay River cascades down the mountains, flows through the town of Ellijay, and joins the Cartecay River to form the Coosawatee River. The Coosawatee spreads out as Carters Lake, a Corps of Engineers dam project completed in 1977 which supposedly controls flooding and provides recreational areas and electrical power for Chattanooga and Atlanta.

Ellijay, from a Cherokee word that means "place of green," was originally referred to in William Bartram's travels as Allagae.

Described by Jimmy Carter in his memoir, *An Outdoor Journal* (Carter 1994), it is where his family had a mountain retreat. Carter spent many summer days as a boy on Turniptown Creek, a small tributary of the Ellijay River. Ellijay's most famous native citizen was Chief Whitepath, a Cherokee who helped Andrew Jackson defeat the Creek Indians at the Battle of Horseshoe Bend. Jackson rewarded him for his efforts by signing the Indian Removal Act of 1830, which resulted in the demise of the Cherokee's homeland. The Cherokee were rounded up, marched to Tennessee, and then forced on to Oklahoma along the infamous Trail of Tears. Chief Whitepath died on that forced march.

Today the Cherokee's homeland consists of cottages strung along the river, new houses on the back roads, cabins and inns in the mountains, and exclusive gated communities. There is even an eighteen-hole golf course named Whitepath in one of these resorts. Its proximity to Atlanta makes it a target for development, and Highway 5 into Ellijay is lined with a Burger King, a Holiday Inn, and chain convenience stores. The new houses in the area are not primary residences; they are weekend getaways affordable only to Atlanta's affluent flatlanders who want mountain retreats. How rich some of us have become to afford second and third homes, vacation homes, and such expensive getaways.

I wonder at the hidden cost of this wealth and whether such wealth is destroying our nation's natural lands. The disturbing fact about Highway 5 is that nearly everywhere I go in this country, I could write the same thing about a similar highway. The phenomenon of land loss to development is happening all over our nation. Suburban sprawl is rampant, not only near megacities like Boston, New York, Chicago, Miami, and Los Angeles, where edge cities have developed like concentric rings of congestion, but all urban areas. For it to be happening in the South seems particularly heartbreaking. The South has always had an intimate relationship with its land and a history of valuing it. It has also had a history of institutionalized exploitation and cruelty, both Indian removal and slavery, but I keep hoping that we Southerners will find better ways of sustaining our economy without destroying our natural lands, without sprawl and development, and without exploiting some of our citizens with low-paying jobs at Wal-Mart. So far, my hopes seem to be failing.

Every year, the making of applesauce in October has been a fall tradition, and we had come to Ellijay for apples. Two weeks after the annual Georgia Apple Festival, the crowds were gone but the apples were still available. We were looking for Mutsu apples. Called Crispins, the variety was created in Japan in 1929 as a cross between the Golden Delicious and a Japanese variety, Indo. Brought to New York in the 40s and then to Gilmer county, the Mutsu apple has become a popular variety in a county which produces over half a million bushels of apples a year.

The ten-mile stretch along Highway 52 known as Apple Alley was dotted with apple houses. There was Aaron's Apple House, Hillcrest Orchards, B.J. Reece's, and many others. To walk into an apple house, any one of the fifteen or twenty in the area, is to experience olfactory overload. The mixtures of sweet and tangy aromas hit us like a warm ocean wave. And the sights were just as lovely. There were baskets overflowing with Golden Delicious and Mutsus—they looked like bushels of gold nuggets—and mounds of Jonathan, Gala, Winesap, Pink Lady, Rome, Arkansas Black, Yates, and Braeburn. Heaped in boxes like rubies in treasure chests, the baskets overflowed with jewels. What a gift John Clayton, who first brought the apple to the region in 1903, gave to all of us who love apples.

Some of the orchards allow you to pick your own. More of a game than a chore, apple picking involves choosing the ones that look and feel the best, filling a bushel basket, and hauling the fruit back to the house. You get so physically involved with the tree and its fruit—the touch of the leaves, the cool breeze, the way you have to coax the apples off the branches—that it's all an intimate and sensual experience.

The history of the apple goes back to the beginning of human time. In Greek mythology, the legend goes, Gaia (mother earth) gave Zeus and Hera a tree of golden apples on their wedding day. Ladon, (the serpent), who never sleeps, guarded the apple tree, and many acts of love, bribery, and temptation take place around those apples. It does not take much imagination to make the leap to the Adam and Eve tale with the apple and the serpent in the Christian creation story of Genesis. Records of the apple's cultivation (Lynd 2006) go back at least 8,000 years to the first human attempts at agriculture in the fertile crescent of the Nile and Tigris-Euphrates, the Indus River valley, and the Yellow River valley. It is not surprising that the making of applesauce is an ancient culinary art. Many recipes have survived, and those from medieval times are remarkably similar to our current ones. The oldest recipe I found on the Internet was one from 1390 (Food Time Line History Notes): "Take apples and poach them. And let them cool and put them though a strainer. And on flesh days add good, rich beef broth and good white grease, and sugar and saffron. On fish days, add almond milk, olive oil and ground spices. And serve it forth." Our recipe was similar except for the white grease. We add a little butter, brown sugar, cinnamon, and nutmeg. And serve it forth. Of course, what makes applesauce is the apples, and the applesauce from those Mutsu apples that year was the best we had ever made.

We make applesauce using a simple method. The apples are washed and quartered into a big pot. With enough water to cover the bottom and on low heat, they simmer down until the whole house smells like a bakery. Friends have been known to drop by and not want to leave. With the occasional stir to keep them from scorching, soon the apples are soft enough to sauce. The apples are then squished through an old-fashioned saucer, a cone-shaped sieve with a wooden cone mallet. That applesaucer is probably seventy-five years old. It belonged to a friend's Dad, who got it from his mother and then passed on to me. Unlike the plastic food processors with their fancy sieve attachments, the old saucer requires a little push work. The apples ooze through the eyes of the strainer, and the pulp is scraped into a bowl. Sugar and spice (and everything nice) add the pizzazz, and the sauce is put into the refrigerator. It's best after a day in the cold.

Our apple-gathering weekend ended with a hike at Amicalola Falls State Park on the access trail leading to Springer Mountain (the southern terminus of the Appalachian Trail). Walking the trail, I thought of the poem by Sidney Lanier called "Thar's More in the Man Than Thar is in the Land" (Lanier 1884). Lanier tells a tale in hill dialect of a man named Jones who could not make a living on his land, and so he sells it off cheaply to a fellow named Brown. He then moves to Texas. Five years later, Jones returns broke and sees what Brown has done. Brown has worked hard, restored the land, planted wheat and corn, and is living prosperously. He invites Jones in for dinner of "vittles smokin' hot" (and I imagine he served luscious applesauce) and tells him "whether men's land was rich or poor thar was more in the man than thar was in the land." These final lines might be interpreted to mean: what makes us prosperous and healthy is the spirit of compassion we extend to the land and to each other; what is in us (and I hope it is compassion and wisdom) will be reflected in our land.

In a way, the poem reflects the reason for making applesauce. We honor the land and those who work to make it productive by transforming its harvest into sustenance. We celebrate the land's gift and give meaning to that gift by making it a tradition. I like to think about that every year when we go about the ritual of making applesauce. In a time when what we eat often comes mass produced (Pollan 2006) with little real appreciation of where it comes from or how it got to be food, picking apples and making applesauce reminds us of our deep and direct connection to our land. The Ellijay River that flows through the apple orchards and defines the watershed of that Georgia land is also what we honor.

Alligator River:
Tidewater Wolves and Swans

shooting star streaked across the sky as we huddled
in our coats, waiting to hear the red wolves. We
were miles from city lights, the air was crisp, and the Milky Way
was a gauze over the night sky. Two planets in opposition, Mars in
the east and Venus in the west, were as bright as a headlight and
a tail-light. The night seemed to be on the brink of balanced
perfection, where anything might occur, even a conversation
with wolves.

The ranger initiated the calls with a single long, sustained
"ahhohhhh." A wolf replied. Then another. Then several voices

joined together in overlapping, undulating howls. The songs ended in a string of yelps. Not many people get to hear the red wolves, the ranger explained, and the Alligator National Wildlife Refuge was one of the few places where the endangered red wolf lived. They were near extinction by 1970 until a recovery effort by the U.S. Fish and Wildlife Service described by Deblieu (1991) brought the wolves back. The refuge is now home to about a hundred wolves. An attempt to reestablish the red wolf in the Great Smoky Mountain National Park (Camuto 2000) had not been successful, but the wolves seem to be thriving in this coastal wildlife refuge.

As I listened to the howls, I thought how wolves had such an undeserved reputation as dangerous creatures; perhaps it was their calls that evoked a sense of danger. The calls were eerie and myste-rious but not really scary. Highly social and complex animals, wolves pose no real threat to humans. Yet our folklore is full of big bad wolves, werewolves, wolves as sheep killers and vicious attackers of the vulnerable human. The wolf has clearly been demonized and considered a scourge. All over the world, wolves have been extermi-nated like no other creature. While such unfounded beliefs were difficult for me to understand, I concluded, like Barry Lopez (1978) in *Of Wolves and Men*, that fear has always found a container for its ugly, irrational nature. Whether a wolf or a person of difference, the container becomes what we fear and despise, a thing to be eradicated.

The home of the red wolf is along the Alligator River, which originates from a lake of the same name. The river twists through the Albemarle Peninsula, swells to lake size, and empties into the sound off the Atlantic Ocean. Bordered on the north by Albemarle Sound and on the south by Pamlico Sound, the peninsula is wrapped by the barrier islands of the Outer Banks. The peninsula's half million acres of wetlands are only a small part of what is known as tidewater country. Although most of the east coast of the United States, from New York's Long Island to the tip of Florida, shares a similar geography—a series of barrier islands that lie off the mainland with inland wetlands—only the inland tidal regions of

Virginia, the Carolinas, and Georgia retain the distinction of the name "tidewater."

North Carolina's Outer Banks are unique in that the wetlands have not been as heavily impacted as in other regions along the east coast. One of the best histories of wetlands, *Discovering the Unknown Landscape: A History of American Wetlands* (Vileisis 1997), traces the loss of nearly half of America's wetlands by the mid-1980s. Although North Carolina's tidewater was drained for agriculture early in our nation's history, these wetlands have remained less developed than other coastal areas. Three centuries of canals, roads, and drainage projects have caused some saltwater intrusion, but efforts like The Nature Conservancy's Albemarle Project to restore the wetlands by closing off drainage routes will help the region to retain much of its original character.

Unlike the condominium coast of Florida or the New Jersey shore with its high population density (1,165 people per square mile, higher than India with 914 or Japan with 835), much of North Carolina's Outer Banks remain undeveloped. With the exception of a few strips of congestion, such as Kitty Hawk, Kill Devil Hills, Nags Head, and Manteo with their rows of three-story vacation beach houses, rental properties, marinas, and shops, the coastline remains salt marshes and swamps. The islands, speckled with a few towns like Corolla, Sanderling, and Duck to the north and Salvo, Avon, Buxton, Hatteras, and Ocracoke to the south, are not extensively urbanized. Inland from the barrier islands, some of the land is cotton and soybean fields surrounding prosperous-looking farmers' houses. Eleven national wildlife refuges and numerous preserves, parks, sanctuaries, and wilderness areas keep the wetlands protected.

Of all the places in the United States, tidewater country has spawned some of the most vivid regional writing of our times. The lines from Stephen Knauth's poem "Testimony" in *The River I Know You By* (1999), "By sepals and pistils we are summoned/to testify against the almighty darkness" astound me every time I read them. In *What the River Means* (1999), Elizabeth Hodges writes of her childhood on the Severn River in Virginia. She describes the river as her teacher: "the river shaped me to fit its curves and bays . . . [it] was the perfect mentor . . . I was the perfect student" (Hodges 1999, 10). The chronicle *Into the Sound Country: A Carolinian's Coastal Plain* (Simpson and Simpson 1997) describes the Simpsons' journeys into the swamps, pine savannas, rivers, lakes, and creeks of Carolina sound country and gives a beautifully written natural history of the region. Outer Banks writer Jan Deblieu, in *Wind: How the Flow of Air Has Shaped Life, Myth, and the Land* (Deblieu 1998), presents a treatise on how wind sculptured the islands and the vegetation. Her narrative, full of lovely phrases such as "I am conscious of the presence of wind because of the voices it gives to the trees," (Deblieu 1999, 110) is one of many that illustrate the magic that tidewater country seems to inspire in its writers.

Tidewater fiction is as exquisite as its nonfiction. Pat Conroy, for example, captures the region like no one else. His intimacy of place is evident in descriptions such as

> the bold, fecund aroma of the tidal marsh, exquisite and sensual, the smell of the South in heat, a smell like new milk, semen, and spilled wine, all perfumed with seawater [and] the perfect coinage of sand dollars, the shapes of flounders inlaid in sand like the silhouettes of ladies in cameos [and] Palmettos close rank at the head of each peninsula and the creek divides into smaller creeks like a vein flowering into capillaries. A sting ray swims just below the surface like a bird in nightmare. The wind lifts off the island, a messenger bearing the odor of moonsage and honeysuckle and jasmine. In an instant the smell of the night changes, recedes, deepens, then recedes again. It is sharp as vinaigrette, singular as bay rum. (Conroy 1986, 5, 4, 400)

There is clearly something about these wetlands that stirs the magic pot of writing.

Earlier in the day, we had driven through the Alligator National Wildlife Refuge and seen large expanses of tidal marshes where cordgrass (*Spartina alterniflora*) and black needle rush (*Juncus roemerianus*) stretched out for miles. Hardwood bottom swamps of cypress (*Taxodium distichum*) lie farther inland, and in between, the unique wetlands called the pocosins connect the watery landscape. Frankenberg (1997) describes these rare wetlands as upland heath bogs full of shrubby evergreens. Located between freshwater marshes and deepwater swamp forest, they are molded by two opposing elements: fire and flooding. From thick layers of accumulated peat, pond pines (*Pinus serotina*) grow in the tall pocosins along with companions such as Atlantic white cedar (*Chamaecyparis thyoides*), swamp bay (*Persea palustris*), swamp magnolia (*Magnolia virginiana*), red maple (*Acer rubrum*), loblolly bay (*Gordonia lasianthus*), and loblolly pine (*Pinus taeda*). In the low pocosins, the taller trees are absent, and the shrubby thickets consist of titi (*Cyrilla racemiflora*), fetterbush (*Lyonia lucida*), sweet gallberry (*Ilex coriacea*), and greenbrier (*Smilax laurifolia*).

The night before the wolf howl, we had gathered in the refuge for an owl prowl; surely the organizer of the events had a sense of tidewater poetic language. The ranger played a tape of a great horned owl and, lo and behold, a great horned owl replied. The bird's outline against the darkening sky caught the last remnants of the day, and as he flew from the middle branches of a loblolly to the top of another pine farther back in the woods, his call was a redundant question, "who, who?" Along another stretch of road, a screech owl answered roll call. Then deeper into the refuge along Milltail Creek, barred owls filled the night with their "who-cooks-for-you?" And it wasn't just one owl. At least half a dozen owls were choraling in some sort of group hoot, and I suddenly understood the expressions *hootenanny* and *hootin' it up*. When one of the birds flew overhead, the ranger shined his flashlight on him. The bird's under wings glowed silver as he soared above our heads. A ghost bird of the swamp, it was the perfect ending for our night of owling.

The next day we were off in search of tundra swans. A fellow at the wolf howl reported seeing six swans on Lake Mattamuskeet, so we drove along Highway 94, crossing the twenty-mile-wide lake slowly in search of the swans. The lake is only about two to three feet deep and was formed centuries ago when fire burned the deep peaty soils. The underground peat smoldered and probably burned for months, causing the land to sink and form a depression. The shallows of Lake Mattamuskeet and nearby Lake Pungo are the wintering grounds of the tundra swans. The birds migrate from their breeding grounds in the remote Canadian boreal forest and tundra west of Hudson Bay. They fly south for thousands of miles, stopping first on the northern Saskatchewan prairies, then on to the Great Lakes and down to the Atlantic coast.

There is no doubt that swans have a fairytale quality. Perhaps it has to do with Hans Christian Andersen's story of the ugly duckling that turned into a beautiful graceful swan. How many of us big ugly girls wished for that story's ending? How many of us hoped and prayed that we would be transformed into such beautiful creatures, only to realize that as adults we would never be accorded that kind of beauty? Or maybe it was seeing the fantasia of Nureyev dancing "Swan Lake." There were so many lovely references to swans. When I recalled the places where I had seen swans—mute swans on Lake Ontario in Braddock Bay, black neck swans on an alpine lake in Patagonia, and whooper swans in the fjords of Iceland—I thought it might have something to do with the places that held these rare and beautiful birds. There was a mystical quality about the places I had seen swans; the landscape and the swans became entwined. All these thoughts drifted through my mind as we searched for the birds on the shallow blue waters of Lake Mattamuskeet.

The ranger at the refuge center confirmed what we suspected. The swans had not yet arrived, and the few that had had not been seen for several days. But the ranger and staff members were so pleasant and so informative about swan biology that it took some of the disappointment out of missing the birds. The ranger told us how their migratory pattern had been worked out using telemetry and how the swan's neck had twenty-five vertebrate, three times the number in a giraffe's neck, and how the birds' bills had razor sharp edges designed to slice off the aquatic grasses they fed on. They told us that of the seven species of swans (whooper, mute, trumpeter, tundra, black, black-neck, Coscorba) only two, the trumpeter and the tundra (once called the whistling swan), were native to North America, although the mute swan had been introduced. The trumpeter swan had been hunted to near-extinction by the 1920s, but conservation efforts brought their numbers back to well over 16,000. During the last eight years, however, nearly 2,000 swans had died from lead poisoning on ponds near the Canadian-U.S. border in Whatcom County, Washington. It was uncertain how the swans were dying, but biologists thought they might be eating lead shot from hunters' pellets or suffering from a bacterial disease similar to botulism. Tundra swans, because they live in more remote areas, have escaped this fate, although the bodies of a few tundra swans had also been found in the Washington ponds. So many things we learned about swans.

Leaving the refuge and driving back over the causeway, I imagined that maybe those white specks far off in the distance were the early tundra swans, but it was too far away to really know. We left unsure if we had actually seen them or not, but I thought that I might return some day to see the 30,000 swans that come in December to winter on the lake. I imagined the sky filled with "chevrons of swans" (Deblieu 1998), their blanket of white settling on the lake like snowfall. That this was their winter home and would be for a long time was the most encouraging thing we learned. Through the stewardship of The Nature Conservancy, the National Wildlife and Fisheries Service, and other public and private organizations, North Carolina's wetlands have not been totally developed, and in preserving these wetlands, they have provided a home for our furred and feathered creatures, a home for wolves and swans.

Yellow River: Conservation, Preservation, and Restoration

The Conecuh Trail that spring morning seemed wildly anticipatory. The pink buds of wild azaleas that laced the path were on the verge of full-blown frill; the cypress pond filled with water lilies was on the brink of bursting into white spikes; common yellow throats flittered about the undergrowth in a low-lying slough; dogwoods splattered white paint spots all over the sandy pine-oak hills. Walking from the trailhead on Highway 137 to Nellie Pond, we were hoping to see the threatened dusky gopher frog. The frog leaves the safety of its burrow (often a gopher tortoise hole, hence its name) and moves to open ponds for the purpose of finding a mate. I wanted to hear the calling song, which has been described as a deeply resonating guttural snore.

I didn't hear the dusky gopher frog, nor did I see a frog that day, but as we walked around Nellie Pond with golden crowned kinglets trilling in the oak branches and a water snake weaving its way along the edge through the pond scum, I was thankful for these trails that gave us access to the beginnings of spring in the piney hills of south Alabama. One of the finest trails in the southeast, the Conecuh Trail, built by the Civilian Conservation Corps, crosses and loops twenty miles through the Conecuh National Forest. The Yellow River borders that national forest.

Although we think of national forests as some of the best examples of conservation in the country, national forests are administered by the U.S. Forest Service, which has a checkered history of conservation both nationally and regionally (Hirt 1994). In the 1930s, the 54,000 acres of Conecuh forest were cut-over, burned-over land. The Conecuh forest was established as a national forest in 1935, and various land management practices have been employed since then. On the positive side, efforts to restore the longleaf pine ecosystem have generally been successful, and many species of the longleaf pine habitat have been saved, if temporarily, from extinction. The dusky gopher frog is one example of a species that is being preserved. The POGO project described in *Pinhook: Finding Wholeness in a Fragmented Land* (Ray 2005) is another valiant conservation effort. This ongoing project consists of acquiring Pinhook Swamp, which connects the Okefenokee Swamp and the Osceola National Forest. When complete, the Okefenokee-Pinhook-Osceola corridor will connect and preserve over 800,000 acres of wild lands. It will be connected to the Gulf Coast Plain Ecosystem Partnership lands (Eglin Air Force Base, Blackwater State Forest, Conecuh National Forest, NW Florida Water Management Area, a Nature Conservancy Preserve, Nokuse Plantation) via Tate's Hell State Forest, Apalachicola National Forest, and Mallory Swamp. The mega-preserve will include several million acres.

On the other hand, the Forest Service has done some irresponsible things. In 1995, after Hurricane Opal, the Forest Service sold off massive amounts of timber to private industry under a controversial bit of legislation called the salvage rider. Using the salvage rider act, the Forest Service sold timber from 50 percent of the forest rather than the 15 percent that had been impacted by the hurricane. Like President Bush's Healthy Forest plan, such policies were not about conservation, but about making money. And the timber industry was the one making the profits off America's public lands.

The Yellow River flows southwest through the Florida Panhandle and empties into Blackwater Bay. Surrounding the river, a basin of swampy hardwood bottomland serves as a protective buffer; it is state land. The Northwest Florida Water Management District has kept much of the basin undeveloped by purchasing 16,000 acres at a cost of $8.3 million. The river, designated as an Outstanding Florida Waterway, is an example of preservation by the state. It is preserved by keeping the land around it free of development.

On our way to the trailhead, we noticed a few signs along Highway 287 around Baker, Florida, that read "Support the Yellow River Reservoir." A Google search revealed the meaning of these signs. A group calling itself Citizens for Water Conservation in Okaloosa County has been promoting the building of a dam on the river since 2001. The dam would create the Yellow River Reservoir (U.S. Water News Online 2002). Some of the members such as Donald Griffith, Mark Weatherly, Ted Mathias and Johnny Johnson (of the Water Conservation Committee of Okaloosa County), Jackie Burkett (the Santa Rosa Country Commissioner and an engineer), Jeff Littrell (director of Okaloosa County Water and Sewer), and others whose names were not included, claim that the dam would provide flood control, hydroelectric power, and a reservoir for much-needed water in the region. They have argued that the dam is necessary for future generations' water needs. They have been accused by environmentalists of having interests in land sales and development around the lake that would be created if the dam were to be built.

Although these individuals comprise a powerful political group, Governor Jeb Bush vetoed $250,000 that was approved by the legislature for planning the dam (St. Petersburg Times, 2001). Those opposed to the damming of the Yellow River seem to be powerful political groups as well. The Friends of the Yellow River include the Audubon Society, the Sierra Club, U.S. Fish and Wildlife Service, Florida Wildlife Federation, NW Florida Water Management District, Florida Wildlife Conservation Commission, the Pensacola Gulf Coastkeepers (a group of concerned citizens who want to preserve the river in its natural state), and several other environmental alliances.

The number of organizations sometimes boggle the mind, so when I met Ernie Rivers, the fellow who first organized Coastkeepers, it was a pleasure. Ernie Rivers is a fisherman extraordinaire, a member of the Bream Fishermen's Association, and a more committed protector of the river you will not find. I heard him speak about Coastkeepers at a Santa Rosa Environmental Defense meeting and was impressed by his passion for the river. He told how he and his group purchased a boat and patrolled the river to ensure its safety much like the original New York Riverkeepers along the Hudson River (Cronin and Kennedy 1997). He explained how he and his fellow members attended county commission and other planning meetings to make sure issues of river conservation were heard and that things did not go on "behind closed doors."

Whether the Friends of the Yellow River and their allies will successfully defend the river and prevent the dam remains to be determined. I hope that what happened in the 1970s with the Tellico dam of eastern Tennessee will not happen here. The story of the Tellico dam, described in *TVA and the Tellico Dam: A Bureaucratic Crisis in Post Industrial America* (Wheeler and McDonald 1986) and updated in a Knoxville zine (Neely 2004), is a classic example of river conflict. The Tellico dam was completed in 1979 on the Little Tennessee River by the Tennessee Valley Authority (TVA) despite the efforts of dispossessed farm families, members of the Cherokee tribe, and environmentalists. The events that occurred pitted the tiny fish known as the snail darter, an endangered species, against the dam in much the same way the spotted owl was positioned against the logging industry in the Pacific Northwest. In their attempts to save the old-growth forests, environmentalists used an endangered species (the spotted owl) in the 1980s. In their attempt to save the Little Tennessee River, the snail darter was used in the 1970s to stop dam construction. Neither attempt was successful, but it slowed both the dam building and the cutting of old-growth forest. Is it any wonder that certain interests want to get rid of the Endangered Species Act? The Tellico dam was built by the TVA despite opposition, and today the area around the Tellico reservoir consists of exclusive resorts, golf courses, and luxury communities with multi-million-dollar homes. Someone made and continues to make a lot of money in real estate in the region.

Whether the Yellow River will remain one of the Florida Panhandle's most beautiful undisturbed rivers or whether it will become another luxury lakefront property, gated communities, or suburban sprawl like that in south Florida or the Disney World sprawl of central Florida, remains to be seen. With St. Joe Company southeast of the Yellow River changing the Florida Panhandle into its *Green Empire* of development (Ziewitz and Wiaz 2004), destruction of the natural environment continues to occur unchecked. But with folks like Ernie Rivers ardently working on behalf of the river, perhaps it will flow free and unobstructed. He claims that it will, as long as he's around, but he's getting on in age.

About a mile north of the river, east of Highway 87, a dry sandy area that once might have been an ancient beachhead stretches out like a dust bowl. The land has almost no trees. It was clear-cut for timber after the Civil War and again in the 1930s. Attempts to farm it failed because the soil was so sandy. Many paved and dirt roads crisscross it like scars. A speedway race track, mobile homes, and some small frame houses stuck in the middle of a bare acre or two characterize the landscape. Some of the trailers are surrounded by junk cars; a huge confederate flag flaps in one front yard. It was a landscape that did not evoke a sense of ease, and I wondered if I was in the wrong place until I pulled into a driveway lined with pines. This was Eleanor's place.

Eleanor had restored ten acres of the cut-over wasteland into healthy Florida mixed scrubland and sandhill. She explained how she bought the old house made of heart pine and moved it right to the middle of her land. Heartwood pine, she said, would never have termites, nor would it rot, because of the natural resins in the wood. The beaded walls were painted off-white, and the floors of six-inch wide planks still had the Baghdad Lumber Mill label on the underside. Hooked rugs covered the floors, an old rocker sat in one corner, and an old pine cabinet nestled in another.

Eleanor said that fifteen years ago there were no trees for miles around, and the wind whipped down like a tornado. Once it was so fierce, it blew the roof off the porch. It sounded like a freight train roaring through, she said. And the heat was so intense during the summer, it burned up everything she planted. It took seven years for the pines (Choctowhatchee sand pine, loblolly initially, then slash and longleafs later) to grow big enough to provide the shade for other plants to grow. Some of the pines now stood forty feet tall and looked like a real forest. She had also planted oaks (live oak, sand oaks, and turkey oaks)—eighteen thousand trees and countless shrubs in all.

We took a stroll around her place. She said she started by planting a perimeter of trees and then made islands. The island mosaics included stands of pine with an open understory of bahia grass (yes, she knew it was an invasive, but it was so hard to remove and the pines would eventually shade it out) and pines with wax myrtle, conradina, calamintha, palmetto, and some other native scrubs. The open meadow had goldenrod, patches of scrub lupine, and other native grasses and annuals. Bluebirds and meadowlarks loved the open meadow. Her bird list had grown from three to several dozen species in fifteen years. Burrowing toads, coachwhips, king and corn snakes were also inhabitants. Near the edge of the open meadow, piles of sand the size of dinner plates made me think of fire ants, but they were pocket gophers. We laughed when she said all she needed now was a barred owl. She said she had recently seen a fox squirrel. She laughed again and said, "If you build it, they'll come."

As we walked in the shade of the young forest among clusters of ferns—probably eight different species in all—and Florida anise, Eleanor spoke of burn ecology. She did not burn. She said burn ecology was not all that good in some places. Some management practices burned every three years, but in reality natural fires were probably far rarer events. Maybe a lightning fire occurred every fifteen years, and while longleaf pines survived fire pretty well, the other pines, oaks, and understory could not. She reminded us of the recent fire in nearby Osceola National Forest that had started as a controlled burn and gotten out of control. Thousand of acres were burned and many homes were threatened in the communities around the forest. As vulnerable as her home was (her old pine house would go up like a tinder box), she was very wary of burns. She thought bush-hogging could probably take the place of burning to maintain an open longleaf ecosystem, but she didn't bush-hog either. She said she cleared the dead limbs and other flammable stuff from under the trees and thought that would probably be enough to prevent fire damage. The discussion turned to turpentine trees, and she related how some old trees, once tapped for turpentine, would explode like dynamite when hit by lightning. Turpentine, once produced all over the Southeast in the 1800s and early 1900s, was a valuable product and provided a living for many folks. She said she knew some old harvest sites where the old trees had "cat claw marks," indicating they were once tapped or "chipped" as Zora Neale Hurston described in her expeditions of the turpentine camps in the 30s.

When it came time to leave, I thought what a pleasure it was to be with someone who knew the history of the land so well and understood how to live on the land in a respectful way. What a joy to be with someone who cared about the forest and the rivers. What a reassurance to be in this small place of restoration. I knew restoration ecology had its critics. A recent article entitled "The Myths of Restoration Ecology" (Hilderbrand et al 2005) outlines some of them, but Eleanor's attempts "to light one small candle rather than curse the dark" seemed to embody the spirit of hope in a region that is posed on its knees awaiting the executioner's sword of developers. Larger efforts to conserve and preserve Gulf Coast lands, such as the private preserve Nokuse Plantation, east of Eglin Air Force Base, will certainly serve to connect the forests of Conecuh, Eglin, Blackwater, then Apalachicola, Osceola, and Okefenokee. They are laudable, but there was something very endearing and hopeful about this tiny place of restoration.

Land and river conservation and restoration are complicated issues. There are really no foolproof ways of protecting land in perpetuity. Even if the government buys land, legislators can sell it off to private interests. The Wilderness Society (2006) reports that Bush's FY07 budget proposes to sell off millions of acres of public land to mining interests and developers. Even if land is put into trust or purchased by a conservation organization, there are still no guarantees. Probably the two best-known organizations for land protection are the Trust for Public Lands and The Nature Conservancy (Brewer 2003). Both are huge organizations that operate in the corporate mode. They have boards of directors whose members have corporate roots. The Trust for Public Lands, established in 1972, has a board of twenty-one directors, many of whom have backgrounds in real estate, development, lumber, and oil businesses. The arguments might be that it takes a big organization to successfully counteract big developers, and who better than the wealthy and the well-connected can successfully fight these powerful interests, but I worry about this argument.

Like the Trust for Public Lands, The Nature Conservancy operates as a large corporation (Birchard 2005). Established in 1951, it employs 3,500 staff members and has more than a million contributing members. It has 146 million acres in projects worldwide and revenues of over $800 million. Its stock holdings are enormous. Its board members and executives are wealthy, well-connected individuals whose backgrounds are in real estate, timber, and resource-utilizing corporations. In short, it is a very rich organization in the corporate model.

There is no doubt that The Nature Conservancy has negotiated deals resulting in the acquisition and protection of enormous amounts of lands by cooperating with big business and government organizations, including the military. The Conservancy's motto, in fact, is "corporate partnership" and it has been called "a real estate deal-making machine" (Birchard 2005). The acquisition of a nearby tract of land called Top Sail Hill just southeast of the Yellow River in Walton County is a good example. The Nature Conservancy purchased this land from a group of developers who had originally bought it from St. Joe Company. Before they could develop it, funds ran out and they declared bankruptcy. The Nature Conservancy quickly purchased it, sold it to the state of Florida as part of the state's Preservation 2000 effort, and now it is a state park. No development will take place on this lovely coastal back dunes ecosystem unless the state legislature enacts some means of selling it off or developing it. Nearby Eglin Air Force Base, with its contiguous corridor of undeveloped land and biodiversity initiatives, is another example of The Nature Conservancy's cooperative relationship with the military (Herring 2005).

The Nature Conservancy has indeed conserved millions of acres of land by negotiating land deals. But still, I worry that an organization that cooperates with big businesses whose own history of land use and resource exploitation has not been good, may not always have the interest of the ordinary citizen or the common good at heart. The fact that the Florida director of land acquisitions for The Nature Conservancy became the corporate vice president for conservation at St. Joe Company in 1999, when this company began its massive development of the Florida Panhandle coast, seems suspect. I am reminded of the fear expressed by a local citizen that summed up the misgivings of those concerned about large, wealthy corporations and big organizations: "I'm against it because when the rich people want it, it's bad for the poor." But The Nature Conservancy may be the only game in town, or at least the only effective way private citizens can protect large amounts of land from being developed.

In a chapter called "Collaborative Conservation" from the book *Keeping Faith with Nature: Ecosystems, Democracy and America Public Lands* (2003), Keiter argues that local grassroots organizations all across the country have been making great headway in land protection by collaborative stewardship efforts involving private and public partnerships. With such homegrown collaborations of local efforts, perhaps the picture is not as bleak as I imagine. How we protect our nation's land and rivers and keep them pristine, or at least healthy, is a hard question to answer. Land use and river health involve complex issues of legality, politics, economics, education, and morality. Issues of conservation, preservation, and restoration are sometimes said to be at odds with one another. But are the issues really so complex? The Yellow River is one of the few rivers in this country never dammed, never extensively developed, and still relatively healthy. The land around it is undeveloped. There is no complexity or ambiguity about that. There is no complexity or ambiguity about my hope that the good people of the region who love the river and its watershed can keep it that way. While some government organizations and large private ones have been successful in protecting our rivers and natural lands, there is something that touches the heart when I think of the individual efforts of people like Ernie Rivers and Eleanor. I wonder if that is where healthy environments begin, with the small and individual acts of love.

PART II

THE AMAZON

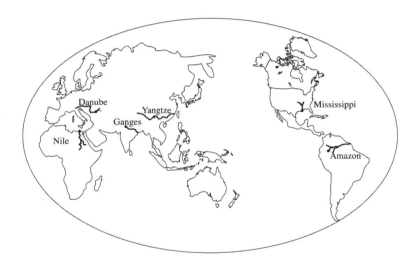

FIGURE 2. Map of the Amazon and other great world rivers

Introduction to the Amazon

To write about a place that so many others have written about—in adventure and travel tales, natural histories, books warning of its demise, scientific treatises and research articles describing newly discovered tribes, animals, plants, fungi, and microorganisms—is daunting. Whether we accept the idea of the Amazon as "A Land of El Dorado," "A Green Hell," "A Vanishing Eden," or any of the other images described in *Entangled Edens: Visions of the Amazon* (Slater 2002), the literature on the Amazon has molded our notion and understanding of it. Both the quality and quantity of that literature is enormous.

Amazon.com, for example, lists over 558 book titles on the subject; Google and Yahoo searches reveal more than 35 million and 8 million Web sites, respectively.

Popular fiction is one of the leading perpetrators of Amazon myths and exaggerations. Pulp fiction, notorious for its sensationalism, is full of Amazonian stories. *Amazonia* (Rollins 2003), for example, begins with a thirty-something fellow wrestling an anaconda and saving an Indian chief's daughter, only to be rescued himself at the last moment from the father's mistaken but vengeful machete. Joining up with a team of CIA agents to find out what really happened to his father in a failed expedition into the Amazon, his adventures roll on like a B movie. Mass-marketed paperbacks often reflect the public's interest in superficial ways and reinforce stereotyped images and myths. Just as those who popularized the American West and created the myth of the cowboy, or those who capitalized on the public's interest in Africa with tales of Tarzan, many fiction writers have used the myth of the Amazon as a hook. In contrast to pulp fiction, novels in the literary genre known as magical realism which mix myth and fantasy with ordinary events and characters, but with far greater wit and sophistication, have sometimes used the Amazon as a setting (Yamashita 1990, Arana 2006).

Portraits of the Amazon in popular nonfiction are not as exaggerated but still have the elements of the bizarre. An article in National Geographic Magazine entitled "Into the Amazon" (Wallace 2003) glorifies the exploits of Indian activist Sidney Possuelo and presents pictures of naked indigenous people, painted and tattooed, with bones or bristles in their noses and ears, bowl-cut hair-dos, and carrying blowguns. The text is wrapped around photos of an enormous anaconda and lush green vegetation muted by torrential rain. National Geographic, NOVA, and other filmmakers present a predictable array of exotic animals, plants, and people of the Amazon regularly on PBS, Discovery, Animal Planet, and other such mainstream television channels. Newspapers and magazines also commonly publish articles on the Amazon that emphasize the extreme and the unusual.

Travelogues and natural histories on the Amazon have been written for well over a hundred years. From the early naturalists (Bates 1862, Wallace 1895, Spruce 1908) to the early travel writers (Champney 1885, Roosevelt 1914), the Amazon has been described as a primitive, dangerous, and exotic place. Nature writers, adventurers, and naturalists from mid-century onward continue to pump out tales and travelogues on the Amazon (Clark 1953, MacCreagh 1961, Shoumatoff 1978, Herzog 1982, Stone 1985, Tidwell 1996, Cahill 1997). Some of the stories are so popular they get republished, as in Roosevelt's adventure retold in the recent best seller, *The River of Doubt: Roosevelt's Darkest Journey* (Millard 2005). In addition to the published works, dozens of personal travelogues appear on the Internet, readily accessible via various search engines.

The scientific literature that comes from the Amazon has also added to its mystique. Indian tribes of the Amazon watershed seem to be a favorite topic for career anthropologists (Levi-Strauss 1955, Chagnon 1983, Ritchie 1995, Tierney 2000). Life scientists, including ethnobotanists, ecologists, biologists, and biochemists, have likewise mined the Amazon for material that is consistently rare and unique. Technical papers reporting new species of plants, insects, and other creatures, new behaviors, new distributions, and new chemicals fill the scientific journals. It would be impossible to cite them all, but an example is the journal, *Emanations*, an undergraduate publication of Cornell University's Esbaran Amazon Field Laboratory, which is located on the Yarapa River, in the region where I traveled.

The abundant pharmacological data have touted rainforest fauna and flora as a panacea for every modern disease from cancer to stroke. I find it oddly incongruous that indigenous Amazonians suffer high mortalities from a plethora of diseases, while we North Americans seek drugs from their habitat to cure our diseases.

Regardless of the irony, ethnobotanists have found the Amazon to be an infinite source of optimistic plant-medicinal information. The popular works of the late director of Harvard's Botanical Museum, Richard Evans Schultes (Schultes and von Reis 1995) and others (Plotkin 1994, Balick et al 1995, Davis 1997), illustrate this scientific publishing phenomenon.

One of the common themes of recent popular and academic scientific publications is that the Amazon is "disappearing" (Smith 1999, Raffles 2002). Clearly, an impacted environment is no myth. In a world where species are disappearing at an alarming rate, the Amazon illustrates this global phenomenon on a large scale. Perhaps of all the literature on the Amazon, these writings are the most relevant. Several publications written specifically for the ecotourist (Lutz 1999, Castner 2000) and the *Smithsonian Atlas of the Amazon* (Goulding et al 2003) were also useful as I traveled the Amazon River.

My trip aboard the riverboat *La Amatista* and my days spent along the Peruvian Amazon River were not unusual. Anybody with a few thousand dollars—maybe even a few hundred—and a valid passport can do it. National Geographic, Audubon, and many eco-adventure companies offer excursions similar to the one I took. Indeed, several million tourists go to the Amazon each year, so my experience was not uncommon. Because the Amazon has been traveled by so many and written about so extensively, the question for me became, "What could I say about it that has not already been said?" Perhaps nothing. But as the saying goes, "No one steps into the same river twice." It is my hope that this narrative, as seen through the eyes of a biologist and a poet, will contribute its particular details and not step into the same river twice.

Of all the rivers on earth, the Amazon most represents our shared biological nature. Metaphorically and in reality, it is life rooted in all its fleshiness, its ingestion and excretion, its growth and reproduction, its breathing. Like the respiratory and circulatory systems that connect the human body with itself, the Amazon connects the earth with itself. Ecologists claim that the river's trees regulate the earth's oxygen cycle; they allow the earth to breath. With its thousand of tributaries, the river resembles the arteries and veins in our bodies. Since the Amazon River has not been impacted by humans as severely as the old-world rivers, it still holds a multitude of natural wonders. It may be the river that offers us the greatest hope, for if we can keep this river ecologically sound and environmentally healthy, we might leave a legacy of long-lasting life rather than devastation. It is with these hopes that I reflect on my time along the Amazon.

Cities of the Amazon

The road from the Iquitos airport to the docks was lined with a long row of wooden shanties and shabby buildings, and brimming with an odd assortment of vehicles, buses, trucks, and motocars. The motocars were three-wheeled chimeras consisting of a motorcycle in front and a buggy with a brightly covered buckboard in back. They buzzed around like wasps. The buses, made of the most readily available material in the rainforest, wood, spewed white smoke as they rumbled down the narrow streets. They reminded me of those old station wagons from the 1950s called woodies.

Through the downtown, where the rubber barons once lived, our minivan joggled towards the docks. One old building, called the iron house, looked like the Eiffel Tower and in fact, according to the guidebook, was actually built by Gustave Eiffel. Others, with their chipped tiles and aging facades, were from the boom-and-bust era when rubber barons ruled the city of Iquitos. The story of rubber (Hobshouse 2003) and the rubber boom (Weinstein 1983) began in the late 1800s. Between 1880 and 1910, rubber was the most valuable commodity in the world, and its demand made some men very rich. It also resulted in cruel and savage extraction practices in Amazonia, including the enslavement of Indians and other peoples. When Indonesia became the leading world producer of rubber by growing rubber trees on plantations, the rubber market in Amazonia collapsed. Iquitos fell into disrepair, and the rubber barons' palaces faded into shabbiness.

As we bounced past the old buildings, I wondered what to make of this faded setting. What to make of a place whose past riches were based on an exploitive economy, and then, when its wealth faded, deteriorated into ruins? The fact that a few men made slaves of others to acquire great wealth brought to mind other times: cotton in the antebellum South, the sisal plantations of the Yucatan, the mines of South America, and on and on. The list could name the times and places in the Americas where the few had brutalized and exploited others for the acquisition of power and wealth. Should we rejoice that those times are gone (but are they really?) and look on these events as lessons? If the lesson is that extravagant wealth is fleeting and that excessive riches in the hands of the few will not last, we have not learned it. We seem unable to break the pattern of greed, and cruelty seems to have no boundaries.

Nearby, Belen floated like a poor relative. Known as the Floating City, more than 80,000 people, mostly of Indian origin, live on this island raft. A large market selling exotic animals and medicinal plants sits near the entrance to the shanty town. Wild animals and plants seemed to be the only things the residents had to sell and one of the few ways of making a living. Of the thousands of houses built on pontoons of balsa rafts, few have electricity, and there is no running water or safe sewage disposal. Everything gets dumped into the river, which is also the source of their drinking water. The waters around Belen are so polluted that no fish can be caught within miles of the city.

I had seen several Web sites promoting Belen as the Venice of the Amazon, but remembered the public health reports of the 1990s. When cholera was emerging in Peru, the town experienced one of the worst outbreaks in South America. Hundreds of people suffered from the water-borne illness. Only the quick mobilization of the country's public health service prevented a death toll of staggering proportions. Having worked with the bacterium *Vibrio cholerae*, I knew the disease to be easily prevented with proper sanitation and easily treated, but with so poor and uneducated a population, the epidemic could have turned Belen into a plague city. I had little desire to see this city of human flotsam up close, even if tourist dollars supposedly enhanced its economy.

Some days later, aboard *La Amatista*, we anchored near the village of Bretana. In the early morning light, I watched thousands of egrets fly across the pink sky. On their way from roosting sites to feeding grounds upriver, they seemed to be pulling daylight across the screen of the sky. As the day lightened, two village women paddled a low-riding dugout across the river from the village to the opposite bank. The river was wide, and with the high water, strong currents and eddies turned the surface into whirlpools. The grace of their strokes and the ease with which they maneuvered their boat through the churning waters was like a dance. Their lean brown arms swung the wooden oars with such choreographed skill, in and out of the red waters. Eventually, they made it across the river, glided into the grass beds, and located their nets. The woman in front lifted the net and pulled in a few fish. They moved a bit further down, lifted the net again and pulled off more fish. Soon their low-riding boat was piled high with fish. They paddled over to our boat, exchanged a few words with one of the crew members, and moved the fish from their boat to ours. They took the cash and paddled back across the river with a small wave of good-bye.

When lunch of fresh catfish was served that day, I wondered if those two women were counting their coins, putting them safely aside, and chatting about what they would do with their money. Would they buy new dresses or new rat traps? A woman we had met earlier had wished for better rat traps. Before harvesting her rice crop, she had to check her traps for rice rats every day, and when she did, she complained about how much damage the rats caused and how poorly the traps worked. Or would these fisherwomen be saving to send their kids to school? Or maybe to buy some new scissors or cooking pots? So many things a woman could do with a little cash.

Another day, the boat stopped at one of the hundreds of smaller villages along the river, and we walked the single dirt street lined with wooden huts. Built ten feet off the ground, the floors were elevated to avoid the high water that came every December through April. Tree trunks formed the support beams, and thatch covered the roofs. The floor planks and half walls were well weathered. They were typical river houses. A generator made electricity for the village, and the lights were on from dark until around ten at night. Chickens and a few pigs roamed freely in the streets, and some wilted red flowers adorned the dirt yards. Behind the houses, a narrow strip of cleared ground was mostly mud. Beneath one hut the fibrous remains of sugar cane lay scattered around a cane press, and under another sat a grinding stone. Bare foot kids in muddy shorts followed us as if we were pied pipers and stared at us with an eerie intensity. At every village and settlement along the river, the kids had that same intensely, curious gaze.

I tried to imagine a life where food production was the main activity and focus of the day. I tried to imagine how hard these river folks worked with limited tools. In all honesty, it was difficult to think of not having a book to read, a glass of iced tea in the hot afternoon, a little music to enjoy, or any one of the many comforts

I take for granted. When we were kids, I remember how we picked pecans, mayhaws, blackberries, strawberries, and figs, how my Dad planted a vegetable garden, and how we ate the fish he caught. But we did not live exclusively off those food sources. The grocery store was right around the corner. Although my grandparents lived primarily off their farms as young men and women and obtained through a system of barter what they needed, how different our world had become in four generations. My brother's kids have no interest in growing food and would think I was crazy to even suggest such a thing. Their time is spent working at jobs accumulating enough money to buy a new car or boat or computer or cell phone or any of the other myriad items available in our world of material plenty. They know little about the natural environment.

The people of these villages, on the other hand, lived in a world of intensive labor and technical scarcity, yet they were intimately connected to their natural environment. They knew the plants and the fish and where to find them. They grew and harvested corn, sugarcane, bananas, yuca (manioc), hearts of palm, beans, breadfruit, and rice. Most farms had orange trees, mango, and other fruit trees. They knew how to build their own huts, cover them with thatched roofs (and replace them every few years) and make a dugout canoe. They knew how to build a water taxi, those large thatch-roofed canoes that moved up and down the river loaded to the brink with bananas.

I wondered if their future would be full of change, if their children would grow up to be accountants and clerks, computer programmers, and technical assistants, and if their cities would be carbon copies of our own. I wondered if progress was always a trade-off. When a culture gained technology, did it lose its connection with the natural world? Maybe the people of these river cities would find a way to have both: a technical world with some of the riches of inventions and comfort-giving goods, as well as one that remained intimately connected to the natural world. I hoped they would not fall prey to other forms of rubber barons.

Amazon Water Lilies

P eople like me, whose ancestors roamed the Scottish moors and the Celtic heath lands, do not really belong in Amazonia. What works best in the Amazon are little and brown. Small and inconspicuous are advantageous qualities in a world full of variation, where bigness is vestigial. Biologically nearly everything has been tried in the rain forest, and big mammals are just another life-form that didn't do especially well and certainly didn't dominate the ecosystem. Even though relatively few big creatures live in the Amazon, 'big' is still the essence of its landscape. No other land mass is as vast and complex as this river basin. To imagine a rain forest three-quarters the size of the United States, to imagine a river nearly a mile wide even at its beginning and nearly 250 miles across where it empties into the Atlantic, to imagine a river 4,000 miles long with 1,110 tributaries, is to understand 'big.' And here we were on the headwaters of the largest river in the world.

We were moving upriver from Iquitos on *La Amatista* towards the heart of the Peruvian Amazon. The night was swashed by the ship's floodlights moving right to left, on and off, illuminating the water full of floating debris. The ship's beam revealed islands of grasses, tree trunks, and limbs. It was the beginning of the high-water season, and the river teemed with flotsam. The summer sun was melting the snow in the Andes, filling the river, eroding the banks, and flooding the lands along the river. Reaching forty-five feet during their peak, the floodwaters would cover much of the land in the floodplain and turn the forests into lakes. Flooded forests are either varzea—the seasonally flooded forests of typical rain forest vegetation—or igapo—the permanently flooded backwater swamps of black water. The uplands above the floodplain that remain dry throughout the year are referred to as terra firme.

High overhead, an occasional lightning bug seemingly the size of a small bird streaked across the black sky. The breeze was cool and laden with moisture. A flash of lightning, then a distant rumbling of thunder mixed with the roar of the boat engine. As the boat chugged along, even in the dark, we could tell the river was very wide. That first night, rocking in the cradle of the river, my dreams were images from my childhood, swimming on the Tickfaw River, fishing on the Blind River, and floating in an inner tube down the Amite River. The next morning, when I opened my cabin shutters for the first glimpse of the river, I was astounded to see a river nearly a mile across and as red as Alabama clay. Leaning against the railing, I watched two tiny brown men paddling their dugout canoe full of plantains upstream against the current. Our diesel-powered boat soon left them behind; our wake gently rolling them back and forth like a rocker.

Later in the morning, we took a small motorboat to explore the backwaters. Zooming along the main channel with our hats flapping against our ears and our speedboat roaring through the rusty waters, we made a quick turn and entered black water. When we entered another side stream, the driver cut the motor and we drifted slowly along a quiet shrub-lined canal. A vast marsh opened before us, and hundreds of barn swallows and white-collared swifts swooped over the grasses in search of insects. White-napped seed-eaters perched on the top of blades of grass and an orange-crowned euphonia stayed only a moment on a bare branch. The driver made another turn, and we entered a shallow pool covered with Amazon water lilies (*Victoria amazonia*).

The lily pads were enormous, six feet across, the size of trampolines! I imagined a small dog walking on them and staying afloat. The edges of the pads curled up, making rims several inches high covered in spikes. The boat shoved up against one of the rims, exposing the underside of the pad and it, too, was covered in sharp spines. Designed to protect the plant from hungry herbivores, the stilettos looked ferocious. The skeletal purple-red underside had the texture of a tin-foil pie plate. A thick stem anchored the pads to the bottom of the pond and a few cantaloupe-sized blossoms the color of buttermilk floated among the pads.

The life cycle of this plant, described in *The Private Lives of Plants* (Attenborough 1995), begins with pollination, an event that might be compared to an orgy. If you consider one of several species of beetle as a young Roman looking to party, the party begins with announcement by the virgin-white flower, which opens during the evening and sends out a delectable scent. The odor has been described as ripe pineapple. At the invitation, the beetle might think, how wonderful, and climb into the blossom. Like any party-animal, he eats and eats, gorging on the special petals inside the flower that are full of sugar and starches, then he dances around and gets covered with pollen. The flower slowly closes, and by morning the beetle is trapped. Maybe the beetle sleeps it off during the day, or maybe the beetle eats, throws up, and eats some more. Who knows with orgies, but by evening of the next day, the flower is purple, its scent is gone, and it opens up, freeing the beetle. The beetle staggers out as if suffering from a hangover, flies away covered with pollen, looking for another party. He won't go to an already-fertilized purple flower, which offers no enticement, but rather seeks another fragrant white blossom. Again he enters the flower, the pollen gets transferred, and fertilization occurs.

From the flower, a spiny fruit the size of a softball develops. Inside the fruit, pea-sized seeds form surrounded by a slimy mucus. If the seeds dry out, they will not germinate. They also have to survive seed-eating fish. Once the seeds fall and germinate in the wet bottom muck, the plant has to grow upward to keep up with the rising waters. And the gigantic pads, the plant's leaves, grow at a phenomenal rate that makes most other plants look like slow Nellies. The leaf grows from fingertip-size to the diameter of a small platter in three days and continues to expand at a rate of six meters in seven days if nutrients are abundant enough, a tremendous amount of leaf biomass to make in a week.

The plant has a remarkable natural history. First discovered by the Bohemian naturalist Thaddaeus Haenke in 1801, during that golden age of naturalists when botanists roamed the world's wildernesses searching for new plants, it remained unknown to the Western world for nearly forty years. It gained notoriety in the 1840s when an English botanist germinated the seeds and grew them in London's Kew Gardens. Causing a stir among the English, whose passion for gardens has bordered on the obsessive if not fanatical, the plant soon appeared in the gardens of royalty, in the gardens of Chatsworth House and Syon House. Joseph Paxton, the gardener for the Duke of Devonshire, obtained a seedling and grew it. The story goes that its remarkable structure inspired Paxton to design a conservatory—in ten days, no less—fashioned after the lily's veined leaves. The conservatory, known as the Crystal Palace, with its transverse iron girders and supports, was the largest glasshouse ever constructed. Built for the Exhibition of 1851, it contained a million square feet of glass, a 120,000-gallon fountain, works of art from around the world, thousands of statues, and life-size replicas of extinct animals. In its gigantic display of English exhibitionism, this Hyde Park marvel—or rather the men who built it—seemed possessed by the same characteristic as the water lily: unbridled expansionism. But after all, that is what life-forms do. Grow. The plant continues to amaze. In 2000 it was featured in an exhibit in the Tropical House of the famous botanical gardens of Oslo known as Victoria House, and it is used today as an attraction in all sorts of garden displays.

The giant water lily is actually one of the few organisms in the Amazon rain forest that is remarkably big. Most of the life-forms of the Amazon, with the exception of a few other plants, are modest in size. One way of thinking about the Amazon rain forest is to consider the phrase, "don't put all your eggs in one basket," or the sound financial advice, "diversify." In some ecosystems, a lot of biomass is invested in large animals, as on the African savanna, where enormous animals such as elephants, giraffes, and buffalo, feed on vast acres of grass. Biomass is heavily concentrated in grass and hoof stock. But the Amazon rain forest has a far more diverse portfolio; it believes in small and many. Putting tiny bits of biomass in its half-million species just seems to be a better investment.

Leaving the quiet pool where the gargantuan flotilla of lilies blanketed the black waters, we returned to *La Amatista*. Motoring back, I thought about what astounding creations come from the artisan river and what endless possibilities the river's watery green hands make—a plant that thrives at the interfaces of soil, water, and air. Living in three worlds, the lily's roots are anchored in the murky substrate of river mud, its stems rise through the dark river waters, and its green pads open to the sky. The Amazon water lily seemed perfectly at home as the trifecta of the river world, a creature of earth, water, and sky.

Birds and Bats of the Amazon

Drifting along a backwater stream one afternoon, I noticed many long hanging baskets on some of the lacy acacia trees. The trees that stood alone out in the open were the ones with most of the baskets. Sometimes as many as three dozen teardrop-shaped bags clustered on one side. If they had been heavy, I imagined the tree tipping over. The baskets were the nests of the oropendolas, the most elegant weavers of the Amazon (Attenborough 1998). Working upside down, the female oropendola wraps a strip of bark around a pronged twig. Additional strips are added in hula-hoop fashion until a skeletal chamber is formed. She

weaves grasses in and out and elongates the stem, in some cases up
to seven feet. The completed nest is a tapestry of beauty and func-
tionality. The long stocking bag, with its tiny stem, prevents snakes
from getting into the nest; its position in clusters out in the open
offers protection from monkeys and other predators.

Continuing along the black-water stream, we passed a cluster
of bushes that jetted out from the bank of the stream. The driver
carefully avoided getting too close to the bushes because hundreds
of wasps buzzed around the foliage. Inside the bushes, the nests of
caciques hung like fuzzy brown tennis balls. Every one of us had
an immediate, instinctive response to avoid that bush. This was
recognition at a primitive level, visual and visceral, a clear warning
of stinging things.

Caciques, like oropendolas, are members of the well-studied
family of birds known as the Icterids (Jaramillo and Burke 1999).
Their nests look like thick balls of twine, and they locate them near
wasp nests. Initially, the wasps sting the birds when they begin to
build and try to drive them away, but after a few days, the wasps
calm down, get used to the birds, and leave them alone. Any predator, however, such as an opossum, a snake, a toucan, or a scaveng-
ing bird that tries to enter the birds' nests, finds itself attacked by
the wasps. Just how the wasps recognize the caciques as opposed to
an intruding foreign bird is not clear, but biologists think the wasps
recognize the strong musty odor of the caciques. It might be that
the wasps smell their benign bird associates and recognize them as
part of home.

Smell can be such a powerful associate of place. One of my
most vivid memories of my childhood home is the aroma of milk-
cake baking. On a cold, rainy afternoon after a long day at school,
that sweet cooked dough brought such comfort and satisfied more
than just physical hunger. Fifty years later, the smell of certain just-
out-of-the-oven cakes can transport me back to that comfort. The
smell of ligustrum is another I associate with home. The heavy
scent is deeply entwined with the comfort of sleep. A large ligustrum

bush bloomed outside my room in summer, and it was especially strong at night just before I drifted off to sleep. Today, when I smell the cloying scent of ligustrum, I find myself yawning. A friend once described how she loved the smell of cooking collard greens. It brought to mind not only her childhood home but her first tiny apartment away from home and her first job. One of the elderly ladies in her building cooked collards every Thursday night, and when she returned from a hard day of work near the end of the week, the smell reminded her of her childhood home when her Mom cooked collards. The familiar smell then became associated with her new home. It was, she claimed, such a complex smell, the layering of childhood safety and adult independence.

Smell can indeed provoke powerful and complex recognitions of places. So it seemed quite understandable that the wasps might come to think of the musty birds as part of their home. An interesting twist in the tale of the wasps and caciques is that the wasps gain from the relationship as well. Certain anteaters and birds such as caracara—and even some species of oropendolas—attack and eat wasps. When these predators appear near the nests in search of wasp morsels, the caciques attack them, with the result that the birds often defend the wasps. Together, birds and wasps have created a sanctuary, a safe haven, and a well-defended home. It is truly an odd coupling of two divergent species, Aves and Insecta, but it seems to work.

Oropendolas belong to the same family as grackles, and they exhibited some of the same bizarre vocalizations and displays. I watched a male oropendola one morning go through a peculiar calling-falling ritual. He threw himself forward almost off the branch, chortling and gulping. He spread his wings out in a flurry to catch himself, pulled his body forward and back upright, and repeated this strange behavior time and time again. He must have performed this maneuver thirty times before he flew off for the branch of another tree to do it all over again.

Another morning, I listened to the babblings of another oropendola in a tree near the riverbank. His noises reminded me of those we made in grade school. I don't know why certain sounds can be so funny, but we entertained each other as kids with our noise-making antics. A Southern comedian once said that his greatest accomplishment was that he could recite the pledge of the allegiance in a single, unbroken belch. My childhood friends weren't that good, but they could make some pretty funny noises. That bird was making some really interesting sounds that morning. He would begin with a few lyrical notes, almost as if he were about to sing a melodious tune, switch to a cranking sort of emanation, and end with a loud plop, almost a gulp, close to the sound of a huge drop of water hitting a pool of still water. I found myself smiling at the bird's song. Why their calls were so amusing is difficult to explain, but perhaps we are still kids at heart and take pleasure in goofy noises. Whatever the reason, I spent a long time listening with delight to the oropendola's vocal ditties and wondered if the humorist P.D.Q. Bach had ever been inspired by the bird.

In contrast to their silly songs and displays, oropendulas are beautiful. The line "She walks in beauty like the night" might well describe them. Their licorice coats take on a multitude of hues. Ornithologists explain this in terms of the physics of feathers. Their feathers act like a prism, and the various colors result not from black pigment but rather from the absorption and refraction of certain lightwaves. Many species of oropendolas also have yellow markings. The flash of a chestnut-headed oropendola as it lifted its hollow-boned body into the air was like a beam of sunlight. When it hit its nest and entered the hole, it glided in as if on a stick of butter. The bird's name seems to suit it, *oro*, meaning "golden," and *pendulum*, from the shape of the nests. With a beak like a knitting needle and a sleek body much like our common grackle, they are as elegant a bird as any in the tropics, yet the locals think of them as common and pay them little mind.

Oropendolas are also in the same family as the brood parasites, the cowbirds. Well known as nuisance birds, cowbirds have increased at the expense of at least a hundred different species of song birds. Birds such as the yellow warblers, song sparrows, chip-ping sparrows, oven birds, yellow throats, red-eyed vireos, eastern phoebes, and eastern towhees are the most common victims that end up raising cowbird young at the expense of their own off-spring. Unlike cowbirds, with their lazy lifestyle, oropendolas are not brood parasites. They spend many hours and much energy feeding and caring for their young.

Of the species of oropendolas (the band-tailed, chestnut-headed, casqued, crested, green, olive, and russet-backed) and caciques (the yellow-rumped, red-rumped, and solitary black) in this region of the Amazon—and we saw them all—my favorite was the chestnut-headed oropendola, with its greenish ivory bill, cocky manner, and noisy shenanigans. I imagine that if Edgar Allan Poe had lived in Amazonia, his most famous poem might have featured this bright-eyed, intelligent creature rapping, tapping, and gurgling at his chamber door, but then his verse might have been more in the style of Ogden Nash than the Poe we know.

One evening I stood on deck watching the river dusk and observed another flyer. The sky had turned into a dome of pastels, and its collage of colors reflected in the water. As the sun dropped below the treetops and evening spread its pink quilt over the river, a hush seemed to fall all around. Suddenly, giant swallows appeared. With silver-tinted wings, they darted over the surface of the water. But they were not birds at all; they were bats, fish-eating bats (*Noctilio leporinus*). They swerved up and down the river, up and over the trees, banked and swirled back down to the surface of the water. Using echolocation to detect the presence of fish, they fished with sound waves. With wing spans of nearly a foot, they were like creatures from the age of dinosaurs. Gliding over the surface of the river, they probed the waters with sonar, trying to detect the minuscule ripples made by the smallest fish. Lowering their claws, they gaffed the fish like ospreys. From out of nowhere, they appeared and kept coming until the sky was filled with black boomerangs.

Because of the droopy jowls that hang on either side of its face, the bat up close looks like a miniature bulldog. In fact, its common name is the bulldog bat, and the jowls are pouches. When a bat gaffs a fish, it stuffs it into its pouches and carries it off. Like a chipmunk stuffing seed into its mouth, the pouches expand, and its shopping-bag function is a convenient way of carrying bat groceries. The bats have large canines, as expected in a flesh-eating animal and their feet are long, with sharp claws. Their bodies look silver because their wings, like the feathers of ducks, have a glistening film of oil that keeps them waterproof. Like all bats, their wings are evolved hands and arms; the bones in their wings are elongated finger bones with a tiny thumb bone (Kurta et al 2005).

I watched the bats swoop over the river, then up over the trees, make spectacular turns, and repeat the maneuver time and time again as they reconnoitered the water for fish ripples. When I was a kid, I loved to watch the purple martins arrive in spring and turn the twilight skies into a circus of aerial acrobatics. They were followed by the summer bats that were even more adept sky pilots. Those insect gatherers turned the evening sky into an exciting place to watch, better than the Blue Angels and without the noise.

The most common mammal in the Amazon, there are about ninety species of bats in the region, and they play a vital role in the ecology of the rain forest. Bats pollinate nearly 70 percent of all the trees in the Amazon. Fruit-eating, or frugivorous, bats disperse seeds and are responsible, in large part, for the regeneration of the forest. A single bat, the short-tailed leaf nose fruit bat, for example, has been reported to disperse 60,000 seeds in one night. Without bats, the rain forest trees would not be pollinated, nor would their seeds be distributed to continue the growth of the forest. In short, the Amazon basin would become treeless without bats.

Earlier in the day as we had motored along a narrow stream, our guide pointed out dark spots on a tree trunk near the water's edge. We drifted over to the trunk and recognized the spots as slumbering bats. Attached to the trunk like refrigerator magnets, these fruit-eating bats were one of the eight common species in the genus *Artibeus* in the region. The sleeping bats were so well camouflaged we could barely make them out. Flat against the tree trunk, they blended into the bark. Hidden from predators such as hawks and owls, they seemed peaceful as they dreamed their little bat dreams. There is something very intimate about watching a creature sleep. They seemed so vulnerable and so like us, breathing in and out, twitching a little in sleepiness. They seemed safe and secure in their sleep, and I wondered if all the world's creatures needed the sanctuary of sleep to keep them safe and healthy.

Bats are often associated with caves, but in the Amazon there are no caves, so bats roost in tree hollows, under large palm leaves, on tree trunks, or almost anywhere that offers shelter. One species of bat collects a leaf, bites it, and bends it around like a blanket. In the daylight hours, they gather for group sleeps, and at night wake to hunt fish or insects, pluck fruit, nibble on leafy vegetation, or sip nectar. The range of bat foods is wide and rather remarkable for a single family of mammals, although there are almost a thousand different species in the family. That an animal could eat so many different foods, sleep in so many different places, and have such a variety of functions seems a wonder, but then bats are a wonder. The only mammal that flies and one of the few creatures of flight whose day is night, bats are nature's odd fellows (Wilson and Tuttle 1997, Findley 1995, Altringham 1998, The Bat Conservation International).

One night, tired and uncomfortable after an especially hot and humid day of birding, I could not fall asleep. I tossed and turned and tried to identify some of the many sounds of the night forest. To get to the toilets required a long, dark walk from my room along a wooden plankway, but I knew sooner or later it had to be done. With my flashlight in hand, I made my way carefully, illuminating my path, stepping up when the walk became elevated and down when the steps declined. The door to the toilets was partially open, so I pushed it forward, and as I did, something dry and leathery brushed against the side of my face. The image that came to mind was a collapsing umbrella. Then nothing. The creature was gone. My reaction was to duck, and when I did, of course, the flashlight rolled off the walkway into the mud. I managed to calm myself, retrieve the flashlight, and get back to the room without screaming my head off. The adrenaline finally wore off and I stopped shaking around dawn. At breakfast, I asked if bats could bump into someone, but our guide laughed and said it was more likely a tropical screech owl. Maybe it was an owl, or maybe just a dream, but somehow I felt that the midnight visitor was a bat, and I had been touched by its hands. Perhaps, if I were born in a different culture, it might have been an omen.

Pacaya Samiria National Preserve of the Amazon

About a hundred miles west of Iquitos, at the convergence of the Maranon and the Ucayala Rivers, the Pacaya Samiria Preserve forms a vast wetlands in the triangle of the rivers. The Peruvian government set aside the region in 1968, and in 1982 UNESCO partnered with The Nature Conservancy to create the national preserve. The largest flooded forest in the world, its five million acres are twice the size of Yellowstone National Park. During the high-water season from February to April, the melting snows from the Andes flow down through the region, forming an enormous sheet of water over the swamplands.

Because the slowly moving waters reflect the images of the trees, it is known as The Forest of Mirrors.

I have spent many days in the marshes and swamps of Louisiana, Georgia, Mississippi, and Florida, but never had I seen anything like these wetlands. Our first boat ride into the preserve (and the only way into it) was an adventure in patience and persistence. The tributaries of the Pacaya River were clogged with floating aquatics, and our driver had to ease through the waters with care. The green mesh that covered the river included two species of water hyacinth (*Eichhorinia* species) and a wild rice (a cultivar from China). Creeping river grass (*Echinochloa polystachya*) and a grass known as capim (*Paspalum* species) created meadows so dense it was difficult to tell river channel from marsh. Making our way slowly through the dense, fibrous mats of vegetation without entangling the motor, we proceeded in starts and stops.

At a bend in the river, the floating vegetation was so thick that our guides decided to hack through a section of flooded forest to take a short-cut around the choked river. With their sharp machetes, they slashed down small trees, sliced off vines, and pushed the boat through the opening. They seemed to relish the path-making, and I wondered if it might have been planned to entertain us visitors. Once, the boat became jammed between two trees, and only after a lot of swinging, chopping, and maneuvering were they able to push it free. Whether or not the hacked alternate route around the clogged river was staged, it provided me a better opportunity to look closely at the infinite variety of plants that inhabited the flooded forest than speeding along the river or walking the treacherous liquid trails.

What a cornucopia of plants the flooded lands held! There were plants with fern-like leaves and lazy fronds, long thin-ridged banners, small leathery chads, succulent tubes, platter-sized, finger-shaped, ceiling-fan-sized, elephant ear, cat's claw, heart-shaped, spade-like, and teacup saucer-shaped, many of them pointing downward like spouts with drip tips, a design that acts as a funnel to allow rain to drain off easily. Lianas (woody vines) added their textures to the jungle mosaic. One called the monkey ladder vine (*Entada gigas*) twisted and curled on either side of the boat like giant rotini. Another among the thousands of species of lianas in this rain forest hung straight down like frayed ropes; others tangled in the branches like discombobulated cables. They were so interwoven with other plants that it was difficult to tell which leaves belonged to what vine or what was supporting tree or shrub. That is what characterizes a rain forest: a tangled web of plants so enmeshed with one another that they are difficult to distinguish. In addition to the infinite variety of leaf shapes, there were fruit and seed pods of all shapes and sizes and taproots that hung from aerial epiphytes. It was a wonderland of water-loving plants. I longed for a field guide of Amazonian plants, but such a book would probably be the size of a refrigerator. While not a field guide, *A Neotropical Companion: An Introduction to the Animals, Plants, and Ecosystem of the New World* (Kricher 1997) is one of the most readable references on the region's flora and fauna.

Finally breaking through the dense tangle, we entered a section of the river free of floating aquatics and made our way to Ranger Station II. Along the river banks, heliconias—there are about a dozen species of *Heliconia* in the area—draped their long, banana-frond leaves over the water. An occasional rack of waxy flowers known as bracts, resembling a string of lobster claws, flashed by as we made our way to the research station.

The research station was a primitive, muddy place. Studies on the threatened side-neck turtles, both the giant South American turtle and the yellow-spotted Amazon River turtle, were conducted here, but it was hard to imagine scientific research going on in this river camp. The turtle-holding tanks were shabby looking containers, and turtle skulls and shells lined the rustic wooden shelves. The cement floors and cinderblock walls were wet and moldy, and there was no sign of anything that resembled a science lab, electricity, or civilization. I felt a growing admiration for those scientists and conservationists who did field work in the Amazon. I could not imagine living in this place for months, weeks, or even a few days, but members of conservation organizations, such as the Peruvian Pro Naturaleza worked here for long periods of time, sometimes years. Their projects of turtle releases and "adopt a turtle" might be what prevent these animals from becoming extinct.

When I thought how these river turtles had nearly been harvested to extinction, it brought back embarrassing memories of my own culpability. As a kid, I used to buy baby turtles from Woolworth's for a couple of bucks and keep them in a terrarium; they always died. Without such dedicated conservationists, both then and now, these turtles might have been hunted to extinction. The conservationists who worked here and, indeed, all over the world are certainly the saviors of our biological heritage. Their dedicated efforts are a redemption from our mindless consumerism. A poem about these modern-day heroes might begin: "Out of the heat and mud of our ignorance, they rescue the earth."

Hiking the three-quarter-mile "nature" trail around the camp, we sloshed through knee-deep water and slick mud. The footing was treacherous, and watching every step along the trail, it was almost impossible to observe the surroundings. The mosquitoes were ferocious; the heat and humidity stifling. By the end of the hike, everyone was covered with mud, thoroughly soaked, and more than a little ready to get back to the boat and into a cooler of drinks. I had never sipped a cold drink that tasted so good.

In the afternoon, we explored some of the preserve's backwaters by boat. Motoring by a forest of aguajales (aguaje) palms (*Mauritia flexuosa*), the trees' trunks looked like telephone poles rising from a mirror; their canopies resembled hairy spiders climbing the sky. Tall, slender trunks with arachnid tops, the trees' reflections in the floodwaters created a duplicate forest. The effect was stunning, like an isometric waterscape. At a river junction, a wood stork stood so still and statue-like, I imagined him a directional sign pointing the way home. Six thousand miles to Florida,

eight thousand miles to New York, it might have read. Farther
upriver, a few solitary cecropia trees rose from the banks. A three-
toed sloth hung from one. Black caiman nestled at the grassy edge,
and capybara wiggled through the water hyacinth, leaving undulating trails as they moved away from the boat.

At several places along the river, we stopped to watch pink
and grey dolphins emerge and submerge. They moved so quickly
we barely glimpsed their humped backs and dorsal fins. One of the
guides called the grey dolphins *tucuxis* and the pink dolphins *botos*.
He explained that the pink dolphins had the ability to stun their
prey (fish, crabs, and turtles) with sound bursts originating from an
organ in their bulging forehead. An account by naturalist Sy
Montgomery (*Journey of the Pink Dolphins: An Amazon Quest* 2000),
describes these remarkable cetaceans. Her writings vividly detail the
myths and stories surrounding them and the conservation work of
Vera da Silva, a leading scientist on pink dolphins, Roxanne Kremer,
a businesswoman-turned-dolphin protector, and others. Montgomery's
encounters with chiggers, mosquitoes, spiders, venomous snakes, and
ants, her descriptions of armored catfish, electric eels, and piranhas,
and her recounting of her travels into the nearby Tamshiyacu-Tahuayo
Community Reserve are some of the most enjoyable and informative
of the Amazon travel adventures written today.

Rounding a curve in the river, someone spotted a monkey
high in the branches of a tree, and the guide identified the monkey
as a brown capuchin and the tree as a canyon ball tree. After a brief
pause, one of our fellow travelers sang out a limerick:

> A brown capuchin monkey in a canyon ball tree,
> trying to crack a nut while scratching at a flea;
> green furry limbs without a single flower,
> the tree's as big as a city water tower.

Something about that monkey evoked such a wave of silliness that
we giggled like school kids. No sooner had the laughter faded when

a troop of black spider monkeys, leaping from limb to limb, had everyone laughing again.

The riverscape kept changing. From tangled jungle to marsh and back again, to an open meadow, then a forest of a single species of palm, then a dense mesh of jungle, then another open meadow, and always below the trees and through the vegetation the slow-moving blanket of water covered the land. One of the guides claimed this flooded orchard was the only place on earth where many species of fish thrived on a diet of fruit and seeds. These fruit-eating fish have been reported to hear the plop of fruit when they hit the water and swim towards the sound like kids to an ice cream truck. Seed-eating fish that defecate seeds, act as plant dispersers and forest regenerators, now that was a remarkable thing to imagine.

We slowed to look for hoatzins in a lazy oxbow lake. From a dense stand of bushes came the sound of rustling leaves and wheezing. Three awkward brown birds the size of large chickens emerged from the greenery. With head feathers standing straight up like punk hairdos, blue-skinned wrinkled necks, and enormous red eyes, it was easy to believe that birds evolved from dinosaurs. These reptilian birds called stink birds, eat leaves that ferment, as in a ruminant's digestive system, and produce a rank odor. Chicks have claws on their wings, and whenever a predator is near, the young birds can hurl themselves out of the nest into the water. When the danger is gone, they use their claws to climb back into the nest. Once the birds fledge, the claws are lost. The endangered birds quickly disappeared back into the dense vegetation, and we set off in search of more sights before the light faded.

On an open stretch of river, a loud chortling scream rose above the roar of the motor, and a bird the size of a goose lifted from the tree tops. "Horned screamer," the guide hollered, and we watched the bird fly off into the distant jungle. As the screamer's cries faded, I noticed chimney swifts zigzagging for insects in the evening sky. The stubby, cigar-shaped birds reminded me of my grandmother's place in Louisiana. One spring, when I was a kid, a colony of swifts, probably just returned from the Amazon, had built nests in her chimney. We heard noises coming from the parlor and walked into a room full of darting birds. Some of the nests had fallen, and the room was buzzing with birds. Grandmother threw open all the windows, took her broom, and wildly swinging it back and forth, swept the swifts from the room. After the birds were gone, she flopped down in her rocker, wiped her forehead with her apron, and said, "Never knew I would be a chimney-swift sweep." Watching the swifts in the Amazon sky, I thought about what an amazing thing it was to see familiar birds in such an exotic place and be reminded of my beloved grandmother.

Night was coming on, and we reluctantly made our way back to *La Amatista*. Storm clouds had settled into the eastern edge of the sky, and bolts of lightning flashed far off in the distance. The rain clouds moved on, the light faded, and the river darkened. The guides pulled out headlights and flashed them along the banks to keep the driver on course. Back and forth the beam moved from bank to bank. Suddenly, the lights washed over two spots the color of hot coals. It was a great potoo! Hunkered down in the crotch of a dead tree, the night bird was an extraordinary discovery. Related to nightjars, goat suckers, frogmouths, and poor wills (wonderful names!), the bird has whiskers like many flycatchers and makes a ghostly call. But this one wasn't making any sound at all, nor was it moving except to close its eyes and flatten itself into the tree until it became part of the trunk.

Above the black tunnel of the river, the sky was illuminated by starlight, and the Milky Way was a ribbon of white sequins. As a closing-of-the-day remark one of the guides said, "The Amazon River is the lungs of the earth," and it seemed a perfect metaphor. These tributaries, like arteries provide nutrients for the body of the rain forest, which in turn, if ecologists are right, allow the whole earth to breathe. The expansion and contraction of the earth with the seasonal movement of water is most extensive in the Amazon. It was certainly not difficult to imagine the earth breathing in and out; it was not hard to imagine all of the earth connected and affected by what went on in this great river and in the Pacaya Samiria Preserve.

PLATE 1. PETROGLYPH AT THREE RIVERS PETROGLYPHS SITE, THREE RIVERS

PLATE 2. GLACIER NATIONAL PARK, FLATHEAD RIVER

PLATE 3. Umbrella magnolia, Cumberland River

PLATE 4. MOOSE, PENOBSCOT RIVER

PLATE 5. REDWOODS, KLAMATH RIVER

Plate 6. Cypress trees, Alligator River

PLATE 7. LONGLEAF PINE FOREST, YELLOW RIVER

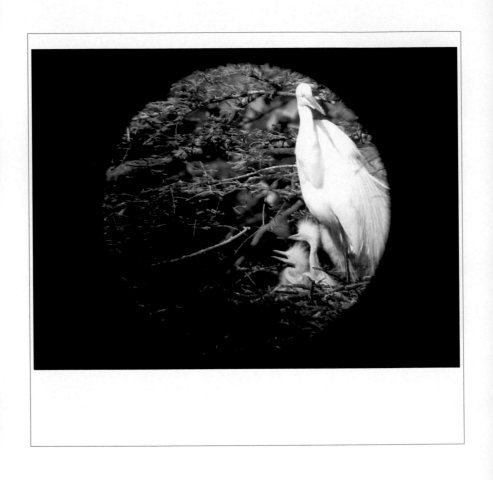

PLATE 8. EGRETS OF CYPRESS ISLANDS, THE MISSISSIPPI DELTA

Plate 9. Amazon water lily, the Amazon River

PLATE 10. DOORWAY OF A TYPICAL RIVERINO HOME, THE AMAZON RIVER

PLATE 11. OROPENDOLA NESTS, THE AMAZON RIVER

PLATE 12. MACAW AT CEIBA TOPS LODGE, THE AMAZON RIVER

PLATE 13. CANAL IN THE HORTOBAGY,
THE DANUBE RIVER WATERSHED

PLATE 14. THE THREE GORGES, THE YANGTZE RIVER

PLATE 15. BRIJ GHAT, THE GANGES RIVER

PLATE 16. THE GREAT SAND SEA, SIWA, EGYPT

Amazon Canopy Walkway

The ride from Iquitos to ExplorNapo Lodge was in a speedboat the locals call a *rapido*; they travel about forty miles an hour. Covered with a canvas top to keep off the sun, with wooden seats arranged like a small bus, they carried about ten people. Zooming along the hundred miles of river, we passed several other rapidos, their sides splashed with big red lettering that read "Explorama." Everyone in the boats waved like friends on a summer outing. We made a quick stop at the Explorama Lodge to drop off a group of tourists staying there and moved on. With the roar of the boat's engine buzzing in my ears, we finally arrived at the ExplorNapo Lodge further upriver. After a light meal of fish and rice and a rest in the canvas hammocks that smelled of mildew, we began our trek to the canopy walkway.

The trail from Napo Lodge to the ACEER field station and then on to the canopy walkway was dark and claustrophobic, and parts of it were under a foot of water. It was like walking through a culvert. The trail was so slick we had several near-falls. But even though we slipped and skidded, we mostly managed to keep our balance. Only once did my traveling companion fall, and when it happened, it was like a pirouette in the dark shadows of the trail. I marveled at the graceful way a person could lose touch with earth so quickly, but I rapidly recanted my thoughts when she came crashing down onto a tree root. Luckily, the root hit in a well-padded region, and she was only slightly bruised. After that, we were both more cautious of our footing, for that forest floor was as treacherous as an icy sidewalk.

The next incident involved a thorn the size of a railroad spike. It went right through one of her sneakers. Luckily, it did not penetrate the skin deeply, but I began imagining all sorts of infections developing. There was something about the heat, humidity, and darkness of that tunnel trail that made me constantly on edge. Images from my old medical microbiology textbooks popped into my mind. Pictures of disfigured feet from chiggers, disintegrated noses and palates from leishmaniasis, crusted-over and blinded eyes from onchocerciasis, awful skin lesions from ecotoparasitic infections like miiasis, and the infinite variety of endoparasitic infections—with names like trichocephaliasis, ancilostomiasis, strongiloidiasis, and belhausia—came flashing back. Thoughts of worms that entered through the soles of the feet, parasites from sandflies that lodged under the skin and ruptured forth as inch-long larvae, black water fevers, and malaria; the list of horrors seemed to grow as we walked along the dark, wet trail. It did not help that mosquitoes buzzed around our ears and would have buried themselves into any exposed body parts except for copious layers of DEET. This was definitely not the kind of hiking that the mountains of Vermont offered. It was not the trails of the Sonora desert or the forests of the Pacific Northwest. This was mud sport.

After what seemed like miles of dark trail, we arrived at the field station, a camp even more primitive, if that was possible, than the lodge. We rested briefly and moved on to the canopy walkway. About ten minutes of walking and we were at the base of the walkway. Metal cables strung from the trees supported a series of suspension bridges constructed of polyester-mesh netting. The structure zigzagged through the forest like a giant woolly caterpillar. Draped from fourteen support trees, the hanging bridges encompassed many acres. About a third of a mile long with twenty-one stations, the wooden platforms were connected by swinging boardwalks made of wire cables, mesh siding, and wooden planks.

The ACEER canopy walkway is billed as the largest treetop walkway in the world. It was built in 1991 as a commercial venture by the two ecotravel companies, Explorama Tours and International Expedition and completed in 1993 at a cost of around a million dollars (Lutz 2000, Castner 1999). ACEER (the Amazon Center for Environmental Education and Research) is a tax exempt nonprofit organization that promotes rain forest conservation, with board members from The Nature Conservancy, National Geographic Society, and several universities (although ACEER is no longer associated with the canopy walk). Explorama started a foundation called CONOPAC (Conservacion de la Naturaleza Amazonica del Peru) which is listed as its Peruvian nonprofit conservation partner but actually consists of Explorama employees. From the various associations, the implications are that the canopy walkway is a research, conservation, educational tool, and it may well be, but Explorama owns exclusive rights to it and uses it as a profit-generating venture. While it may host scientists, provide educational programs, and be "dedicated to rain forest conservation," as its advertisements claim, it functions primarily as an ecotourist attraction.

Ecotourist attractions and ecotourism in general have raised many concerns of truth-in-labeling and "who owns nature" issues.

Promotions of tours with buzzwords like *green travel, ecotravel* or *nature travel* have become common. Travel companies and ecolodges that claim to create a sustainable economy, promote community-based conservation, and partnering with conservation groups represent marketing and profit-generating strategies. The tactic known as "green washing" is a growing phenomenon (Honey 1999, The International Ecotourism Society). Whether Explorama and its canopy walkway were part of a truly successful effort to improve the lives of the people of the region, as it claims to do, I don't know. I also don't know how the money flows. I learned after my travels that International Expedition had merged with seven other travel companies to form Grand Expeditions, Inc., which was then acquired by First Choice Holidays PLC, which made me wonder about local involvement. First Choice, listed on the stock exchange as FCD.L, owns Royal Caribbean Cruises and Island Cruises, among other holdings, had a 2005 net profit of $145 million and a typical corporate structure: wealthy chairman of the board, high-salaried CEO, and many high-salaried directors. The canopy walkway did not feel like a tourist attraction, but it may be controlled by corporate interests, and the dilemma is that it was the only means I could find to see the rain forest canopy of the region.

Up the wooden ladder and onto the first wooden platform we climbed. The first steps on the walkway took a bit of adjustment before I learned to relax my knees and let my legs sway with the movement of the planks. Our guide asked us not to touch the wires because the insect repellant on our hands would dissolve the polyester fibers, so walking was a balancing act. Wobbling along the suspension bridges, I noticed the forest floor was not flat. The walkway ran along the top of a ridge and across a ravine, so there were many different levels of ground beneath us. How did they build this spider web in such terrain? I wondered.

As we climbed, the darkness of the forest floor lifted, and sunlight brightened everything around us. The higher we climbed, the cooler and lighter it became. The transformation happened quickly, as a day suddenly brightens after a thunderstorm. The dark shadows of the forest floor dissolved and a brighter green took its place. The colors reminded me of my early tree-climbing days. My favorite tree was an old camphor tree that offered thick, comfortable limbs and a bright green hue. The shades of green in the rain forest canopy were the same as in that old camphor tree, and it all felt familiar. I thought, how lucky canopy biologists were to relive their childhood tree-climbing days each day they went to work.

The first birds we saw were two white-necked puffbirds tucked in a branch close to the trunk of the tree. Nestled in a green nook with their black and white feathers fluffed out, they looked like an elderly Russian couple in winter coats. Why I kept conjuring up images that had to do with cold climate, I do not know. Perhaps it was just wishful thinking. Off to one side, a bromeliad the size of a walk-in closet sprouted from the crotch of a tree. A bromeliad's aerial microcosm can support thousands of inhabitants, from invertebrates to insects to frogs, and I marveled at the size of that leafy valley metropolis.

From platform six we climbed up two more levels and came to the highest point on the walkway. It was over 120 feet high, the height of a twelve-story building. The breeze was cool and the view from the green aerie was stunning. Trees stretched out unbroken like rolling hills as far as I could see: nothing but trees, no clearing, no break, no sign of humans whatsoever. A misty halo above the treetops gave the forest a foggy look. The mist was due to evaporates which form around the tree crowns. Rain forest trees act as massive water pumps, and millions of gallons of water from the wet soil get transported up to the leaves, where some of it evaporates. Just above the trees, the evaporates form a mist that hovers like a low-ceiling cloud. I thought the rain forest trees were living with their heads in the clouds.

Moving to the next platform, we observed an arboreal lizard hanging onto the trunk of a tornillo tree and feeding on ants. The lizard, known as the beaver-tailed lizard, was sucking up ants like a vacuum cleaner. With its speckled skin, two-ring collar, and thick spiny tail, it resembled a Gila monster. Its long delicate toes, tipped with sharp nails, gripped the trunk as it scampered up the tree. On a philodendron, a firestick, one of many species of walking sticks in the region, beamed warning lines in stop-sign red and mimicked the movement of a twig.

At the next platform, we paused to watch tanagers fidgeting in the treetops and toucans preening on a distant branch. A flock of green-headed parrots flew overhead like apostrophes. Rarely seen other than in pairs, they filled the sky with a congregation of quotes. A flock of cobalt-winged parakeets bobbed overhead, and red-bellied macaws sailed smoothly across the sky. A spangled cotinga sparkled in the branches like a jewel pin, a ruby mounted in turquoise.

Back down on the forest floor, it took some adjustment to get used to the dark forest again. The boat ride back to the lodge along a tiny stream was a tight squeeze as the driver maneuvered the boat through the narrow channel. Five minutes into the ride, the rains came, not as gentle showers but in torrents. In all my days, I had never encountered so much rain so quickly. In thirty seconds there was an inch of water in the boat; in a minute there were several inches. It was a reminder of the nature of this forest. It rains over a hundred inches annually and 200 days out of the year. They do not call it a rain forest for nothing. We sat huddled on the seat with ponchos over our shoulders and tarps drooped over our belongings. Within minutes we were drenched. Our duffel bags were drenched, our boots were full of water, and we sloshed when we stood up and walked. We learned that nothing ever really dries in the Amazon, and we never had completely dry clothes again until we returned home. It was a place that embraced the overwhelming physical reality of water.

Insects of the Amazon

Our guide, Ricardo, was intent on finding the three species of manakins—wire-tailed, white-crowned, and golden-headed—that lived near Napo Lodge. A good guide, he wanted us to see these rare and beautiful birds. Since they were strongly territorial birds, they should be easy to find, he explained, so we followed him like ducklings along the trail. But when he veered off into the dense underbrush with his machete swinging like a pendulum, I began to worry. The marked trail had offered a certain degree of security; it was open, visibility was better, and the idea that other humans had walked it before and left a human

scent made me feel a little safer. When Ricardo took to the underbrush, I began imagining an encounter with the deadly fer-de-
lance, the most common venomous snake in Amazonia, or finding
one of the thousands of species of biting insects crawling up my
pant leg, looking for lunch.

Taking a deep breath and saying a silent prayer, I plunged
into the underbrush after him. The smell of damp, moldy earth
was strong through the newly cut paths. Between swings, he
searched the trees for the manakins. Several hundred feet off the
trail and concentrating on the treetops, the ground beneath our feet
suddenly grew wavy. Ricardo shouted, "Army ants. Give them
room!" They came streaming up from behind us in fast moving
columns that branched out like streams of water trying to find the
path of least resistance. They were foraging, and anything and
everything, it seemed, was fair game. As they moved forward, all
sorts of insects and tiny animals jumped up, trying to either outrun
the pack or get out of their way. Following the ants were the ant
birds, scooping up insects that the pack had disturbed.

Hopping over a stream of ants and out of their path, I was
astounded to see how many ants were moving; generally colonies of
army ants averaged about 100,000 individuals, with foraging par-
ties of around 10,000. The hordes poured over the ground like
black oil. While each ant was only about a quarter-inch long, collectively they made long, rapidly moving lines. I was reminded of
a miniature stampede, the sort of mass movement wildebeest make
when they rumble over the Serengeti plains during their annual
migration. And it turns out that these ants have a lot in common
with wildebeest. Their whole existence is as part of a herd. They
even have herd followers that prey on them. Beetles called sta-
phylinids follow the ants and prey on the weak ones, the same way
lions, cheetahs, and hyenas do on wildebeest. To add to the complexity, there are camp-following beetles known as linulodids that
ride on the backs of the worker ants. In some species of army ants,
a silverfish species feeds on their secretions and even shares their

prey. Unlike wildebeest, which are herbivores, army ants are carnivores that group-hunt for prey. While the ants are raiding and foraging, predators are culling their herd and preying on them. Meanwhile other insects are riding their backs and sharing their food. Ant birds—more than forty-five species in the region—follow them around, feeding on the insects they flush. It is a very hectic world down there on the jungle floor.

Army ants—there are about twelve species in the genus, *Eciton*, common in the area—have the most complex organizational behavior of all social insects with the exception of honey bees. Living cyclic nomadic and stationary lives, they are both wanderers and homebodies. They bivouac in large aggregates the individual worker ants create by attaching themselves to each other. Heads down, they interlock their tarsal claws to create a lattice of ant bodies, giving literal accuracy to the term *homebodies*. Like any densely populated city, of course, they produce garbage, and their refuse dumps, or middens, are commonly found near the colony. They have a queen as the reproductive specialist and generally two female castes, workers and soldiers (Gotwald 1995, Holldobler and Wilson 1990).

In our search for the manakins, Ricardo pointed out another type of ant and warned us to avoid it. A large dark solitary ant was making its way up the trunk of a cecropia tree. About an inch long, the bala or bullet ant (*Paraponera clavata*) was hunting. These ants live in the ground near the base of tree trunks and are predators, scavengers, and collectors of nectar. Hoyt (1996) tells the story of two bullet ants to illustrate the complex relationships between these ants and the balsa tree. The ant sisters detect the intoxicating scent of the tree and feed not on the flowers but on special sugar pots called extrafloral nectaries around the petiole of the leaf. The nectar is high in carbohydrates but not amino acids, so in order to get the essential amino acids, the ants eat pest insects on the tree for the additional protein. The balsa tree provides nectar for the ant while the ants reciprocate by eliminating pests that might damage the tree.

Not especially aggressive, bala ants will sting only if disturbed. Ricardo told us of a student working in the area who accidentally leaned against a tree without looking. His hand landed right on top of a bullet ant, and he was stung. The wound became swollen and very painful. For two weeks the fellow was quite sick. What made him ill was a neurotoxin called poneratoxin, which functions as an insecticide. It damages the nervous system of insects, but its effect on humans can be serious as well.

On the way back to the lodge, we noticed another line of ants moving across the trail carrying leafy green cargo. We stepped gingerly over the leaf-cutting ants with respect. How could you not admire these hard-working gardeners? They are the quintessential tillers of the rain forest and cultivators of fungi (Taber 1998). Like all ants (Gordon 1999), they live socially complex lives, communicating with one another using an array of chemicals, but these ants specialize in harvesting and carrying enormous amounts of greenery to their nest. Like little winemakers, they cultivate fungi on the collected harvest via a fermentation process and use the fungi as food. Remarkably, the fungi can even communicate with the ants. If a material brought in is toxic to the fungi, the fungi release a chemical that tells the ants to stop collecting that particular plant.

A tale of these sauba ants (*Atta* species) is told in the adventure tale, *White Waters and Black* (MacCreagh 1961). MacCreagh's exploration party had stopped along the river at a deserted community house. The ground was high, and it seemed to be a good place to camp for the night, so they slung their hammocks up using the old roof poles for support. During the night, they heard faint clicking noises. Turning on the flashlight, they couldn't see anything unusual. But in the morning they discovered why the place was deserted. Sauba ants had come in the night and left all their clothes in shreds. MacCreagh's khaki shirt was a handful of rags. It looked like what a hungry caterpillar would do to a tasty leaf. His trousers were completely eaten except for the buttons, and his boots were digested down to the soles. The clicking noise had been the sound of the ants' jaws working away, chewing all night on human's clothes.

Another vivid description of an encounter with these ants comes from the botanist David Campbell (2005), who calls them sauva ants. He describes arriving at one of his forest sites and being overwhelmed by the smell of formic acid. The ants had dribbled formic acid everywhere to mark their trails, and after a rain the air was saturated with the caustic, sour odor.

The common occurrence of these ants in indigenous people's creation stories as Harvest Mother and other goddesses (Lauck 2000) illustrates the vital part these ants have in their world view and mythology. The women of the Amazonian tribe known as the Kayapo consider ants the guardians of their fields, friends, and relatives. The women paint ant designs on their faces and understand that the ants keep certain vines from choking out the manioc, or cassava, one of their important food plants.

Of the nearly nine thousand species of ants in the world, many hundreds are native to this region of the Amazon. In fact, there are more ants than any other creature in the Amazon. Our day was intended to be a birding day, and we did see the three species of manakins. Brightly colored bangles, they were a joy to observe. But our encounters with the ants reminded me that on a relative scale, the world is mostly full of small, dull-colored creatures; as Holldobler and Wilson (1990) write, these diminutive clones "run much of the terrestrial world . . ."

That night, as I lay on the thin, damp mattress, I wondered why on earth I had spent so much money to sleep in a six-by-six foot bare cubicle with no electricity, no flush toilet, no running water, and raw plank walls open at the top so that every sound a guest made was shared by all. The sounds of the night jungle surrounded me. Every bit as exotic as those new-age tapes of rain forests with eerie bird calls, falling rain, the clicking and shrilling of insects, the hum of katydids, and the bufoing of frogs, the forest was a rhythm section in a samba band. But the noise of this nocturne was anything but soothing. What can seem exotic and intriguing in a familiar, cool, dry climate can be really annoying when one is exhausted and sleepless in heavy humidity and intense heat.

A Louisiana fish camp might look luxurious by comparison. Lukewarm water in common makeshift showers, plain food, odorous pit toilets, and dark walkways lit by kerosene lanterns that emitted black plumes of smoke: it was not a pleasure dome. A blanket of heat and humidity covered me as I lay in the white shroud of the mosquito net. Too hot to sleep, I shifted from one position to another. To find some distraction, I scanned the ceiling and noticed several dots of light on the thatched roof. In the dark steeple of the roof about twenty feet above my cubicle, three greenish spots glowed. Maybe it was a reflection, or stars shining through holes in the thatch, but as I moved my eyes slightly off center, letting those light-detecting rod cells do their job, I realized that it was light. Two greenish-yellow spots side by side and a third below would fade and then reappear. The next morning, I learned from our guide that the spots probably belonged to a beetle known as the fire beetle, headlight beetle, ghost eyes, or click beetle, a bioluminescent insect.

The fireflies of North American summer nights are familiar bioluminescent insects. Although in recent years I've seen them less frequently, I discovered that more than two-thousand species of insects produce light. (A good review of bioluminescence is by Babu and Kannan [2002] and an overview of color can be found in *The Seven Deadly Colours* [Parker 2005]). Even though I had a hard time seeing the faint light produced by the insect on the roof, these beetles (*Pyrophorus* species) produce some of the brightest light in the world. Found in tropical regions such as Jamaica, Indonesia, Costa Rica, and the Amazon rain forest, they produce light in special cells called photocytes. Their light shines through a translucent cuticle, and a reflecting surface (much like a mirror) behind the photocytes helps intensify it.

When I consider the miracle of light production, I am awed that creatures make light. I know that scientists explain bioluminescence in many ways. Microbiologists and biochemists who first studied bioluminescent bacteria describe it as a chemical reaction: in the presence of the enzyme luciferinase and the substrate luciferin, oxygen and the energy molecule ATP result in light. All of these compounds are coded in the genes; some light is green and some is yellow or blue or red. According to molecular biologists, a different codon—those three nucleotides in the DNA that form the basis of the genetic language—or a single difference in the amino acid sequence in the enzyme luciferinase results in a different color of light. In other words, one amino acid can make the difference between green and yellow bioluminescence.

Physicists explain bioluminescence in terms of various wavelengths. Green light, at 492 to 577 nanometers, consists of many shades of green, such as ordinary green (at 545 nm), watermelon-rind green (around 500 nm), and lime-Jell-O green (around 560 nm). Yellow light, on the other hand, has a longer wavelength of 597 to 577 nm, while blue has a shorter wavelength in the 455 to 492 nm range, and so forth. Some theoretical physicists claim that the whole thing consists of tiny vibrating strands of energy. Animal behaviorists and physiologists describe light production as a means of attracting mates or prey or to ward off predators. Evolutionary biologists describe light production as a genetic trait that increased reproductive success. There are so many ways of describing light.

Of all the explanations, I like the concept of light as a language best. I began to imagine the transformation of light into our languages. I imagined Spanish as various shades of red and Russian as blues. I imagined verbs as the most intense and nouns nearly so. Adjectives and adverbs would have the least amount of brightness, and pronouns would be rather dim. But how could shades and intensities of light possibly match the complexities needed for language? I may as well have asked how four nucleotides could encode all the genetic complexity of life?

We "higher animals" are so used to thinking of communication as mediated by sound and/or light because we make noise and see mostly visual clues from reflected light. Humans have developed sign languages as an alternative to vocal language, but there are so many other ways creatures communicate. In biology departments, there are entire courses called "Communication" with topics that vary from plants that beckon pollinators in the language of fragrance to plants that sweet-talk bacteria into moving closer with attractants called lectins, bees who communicate with each other by dance and learn the location of food through gyrations, and of course, the social insects such as ants that communicate by acidic chemicals and touch.

How varied the concept of vision and sound can be as ways of communicating among organisms. Many pollinators see flowers differently than we do. A drab white flower can be a dazzling spectacle when seen through the eyes of certain insects that detect light waves of different frequencies than our own. We cannot hear the night whines of bats at all. It is all a matter of perception. All those sound waves and light waves were energies that could be transformed one into another. How easily we do this with our electronic gizmos, like those little screens on stereos that bounce with multicolored bars when certain notes are played, the way TVs take electrical impulses and transform them into images, and all the multitude of recording devices. Sound is transformed into electrical waves, then into light, and back again into sound. Was it all interrelated, sound and light and language?

As my eyes grew heavy, the night seemed full of language. The sounds of creatures I could hear, the beetle's luminescent sentences I could faintly see; as a sweet odor drifted in, I recognized the lovely language of smell. All those chemicals floating in the air, plants signaling plants, insects calling insects, flowers beckoning insects, all the myriad chemicals that connect creatures. Every living creature appears to need to communicate with its own and other species in some fashion or other. Maybe this need for language is as essential as food. My last thoughts as I drifted off into sleep were to wonder if there were unknown languages, secret languages, languages that were transmitted in ways other than through sound or light or chemicals, languages we have yet to discover. How many wondrous ways do we, the creatures of the earth, have to make conversation?

Trees of the Amazon

We spent our last days along the Amazon River in a lodge called Ceiba Tops. Our room had air-conditioning, a flush toilet, and a hot shower. What luxury, sleeping in a cool room that did not smell sour and showering with hot water after a day in the jungle. But it was not just the creature comforts that made it enjoyable. It was the brightness of the place that lifted my spirits. Although trees surrounded the lodge, the place was more open, located right on the river, and sunlight poured in. Unlike the dark Napo Lodge, Ceiba Tops was bright and cheerful. It was odd that fewer trees created a more pleasant surrounding, because usually the more trees, the merrier.

Ceiba Tops' namesake is the ceiba tree (*Ceiba pentandra*), probably the most representative tree of the tropical rain forest (Prance
and Lovejoy 1985). This icon is the tallest tree of the rain forest
reaching heights of two hundred feet. With flared buttresses and
widths of twelve feet, it has the perfect design to support an enormous canopy. These arboreal giants can live to be five hundred
years old, and the creatures that make their homes in them number
in the thousands. E. O. Wilson reported finding forty-three species
of ants living on a single ceiba tree (Kricher 1997). Not only do
epiphytes such as ferns, orchids, bromeliads, lianas, and vines grow
on the branches, but bats, frogs, spiders, and snakes live in their
cavities, insects reside all over the trees, and sloths, monkeys, and
birds inhabit the leafy canopy. In one ancient tree, even the remains
of a primitive human camp were found in its buttress.

In the same family as the baobab of Africa, *C. pentandra* is
called the kapok tree because of the silky cotton floss, or kapok, it
produces. The seed's kapok wings, form a football-sized pod, and
when it ripens and bursts open, the kapok drifts in the air like
snowflakes. Too brittle to weave into fabric, kapok fiber has the
properties of a water repellent and has been used as stuffing in life
jackets and seat cushions. The leaves of the tree are drought deciduous, which means they fall off when there is no rain. The tree's
white-pink flower attracts many floral visitors and is pollinated by
two species of spear-nose bats. Its soft wood is unsuitable for lumber, but it is cut for pulp and trash wood. Ceiba trees are also cut
and made into framing for concrete molds. When the cement hard-
ens, the wood is discarded and left to rot. I tried to imagine my
house being torn down and used to prop up a sidewalk and then
thrown away.

Ecologists claim the Amazon River ecosystem has two-thirds
of all its carbon tied up in trees. David Campbell, a botanist who
has studied tropical forests for over thirty years, reports that in his
forty-four acre research site near the Peru-Brazil border, the trees
numbered over 20,000 and belonged to more than 2,000 species

(Campbell 2005). This is the largest number of trees in any ecosystem on earth and over three times the tree species in all of North America. It is not surprising that much of the natural history of the rain forest involves trees and that the world knows the Amazon by its trees.

Probably the second most recognizable rain forest tree is the cecropia tree; there are more than a hundred species of *Cecropia*. Easily identified by its rosette leaf pattern, the cecropia is considered a secondary tree because, although it is a fast grower, it does not usually get as tall as the ceiba tree. Like a Hindu goddess with her many arms holding up hundreds of ruffled parasols, the tree sprawls along the open areas of the riverbanks. The tree requires a lot of sunlight. Cecropia is well studied by ecologists because of its mutualistic relationship with Aztec ants, but its largest occupant is the three-toed sloth. The sloth is almost always found in the cepropia's branches, camouflaged in a coat of green algae and munching cepropia leaves. Cecropia fruits also provide an important food source, with as many as forty-eight different species of animals and birds eating them.

Palm trees are also common, and there are hundreds of species. The keystone species of the region is the tall aguaje (*Maurita flexuosa*). Pulped and added to popsicles or ice cream, dried and made into flour, or fermented and used as a drink, their fruits are probably the most economically valuable as a local food source. A smaller palm, the thatch palm (*Lepidocaryum tenue*) is the one used most commonly for making thatch roofs and coverings. Being lightweight, waterproof, and porous, the fronds are ideal for the rain forest climate.

One morning, Ricardo pointed out some of the sap-producing trees along the trail. The most infamous sap producer, the rubber tree (*Hevea brasiliensis*) was a rather ordinary-looking tree for all its notoriety. A member of the spurge family, it produces raw rubber, which is harvested much like maple syrup. A slash is made in the trunk's smooth bark, and the milky latex drips into cups. The latex drippings are rolled into balls and sold for commercial processing. Latex, a polymer in the polyterpene family, is toxic to some insects, so it functions for the tree as a deterrent against insect predators.

Another sap-producing tree, the chicle tree (*Manilkara zapota*) has as its claim to fame the making of Chicklet gum. By adding sugar to its soft chewable sap, chewing gum was invented. Yet another sap-maker, the bleeding tree or dragon's blood tree (*Dracaena draco*) produces a red sap. To demonstrate, Ricardo sliced a two-inch gash across its smooth trunk, and a transparent red fluid oozed from the cut. He said the sap contained compounds that aided in healing wounds. The most interesting sap tree was the possum wood tree (*Hura crepitans*). According to local fishermen, wood storks use the tree to produce fish kills. They tell the story of seeing wood storks poke holes in the tree and rub their bodies around the cut to coat their feathers with the sap. The birds then rinsed in the river, and the sap, which contains a powerful neurotoxin, poisoned nearby fish. When the fish died and floated to the surface, the storks made a meal of their kill. How about that for "better living through chemistry"?

One morning, leaving Napo Lodge, as our boat curved around the bend of the Sucusari River and sped into the Napo River, we encountered a tugboat with several rather pitiful-looking characters aboard. They were loggers collecting the tree trunks tied together and floating near the bank. The three men on deck wore ragged shorts, T-shirts, and flip-flops and looked like skeletons as they squatted over a metal pot that contained a fire. Whatever they were cooking, it appeared meager. They glanced furtively in our direction but unlike most river people, neither waved nor smiled. I had the impression they knew that we knew their activities were illegal. I don't know what I expected, but I could not imagine those poor emaciated men as loggers. Perhaps I thought loggers would be fat, lazy, greedy, and arrogant, but these men looked so desperately poor that it was hard to work up a dislike for them. Instead, their appearance evoked sympathy.

Speeding around their boat, I thought, how often the truth lies in complex wrappings, and when we unravel one issue that involves exploitation there are often multiple levels of exploitation going on. I learned that these men were poor farmers who came down from the mountains to earn a little money the only way they could, by logging. They cut and sold cedar (*Cedrela odorata*) and mahogany (*Swietenia macrophylla*, an endangered species) to brokers who sold them to timber companies, who then sold them on the international market. The same log that a logger might sell for $30 would bring about $128,000 as furniture. So the exploitation was not only of trees but displaced farmers as well. Who were these companies that hire these displaced farmers? They are wealthy businessmen who have created companies that provide a safe distance from the consequences of their actions. So often this is the way exploitation works. An organization is created where the many layers of transactions make it difficult to hold accountable those responsible.

Near the mouth of the Amazon River in Anapu, Brazil, the battle for trees has resulted in bitter conflicts that escalated into violence with the brutal shooting of a beloved nun, Dorothy Stang (Downie, 2005). The incident grew out of tensions between poor residents and logging companies, ranchers, and land speculators, in other words between the poor and the rich. Reports of killings of hundreds of rural workers, landless farmers, slave laborers, and activists went unnoticed until the well-known nun was murdered. When the Brazilian government rescinded legislation requiring logging licenses and allowed logging to continue, the violence escalated and troops had to be employed to curb the tensions. In April 2006, a little over a year after the killing, Amair Feijoli da Cunha was convicted of hiring two gunmen, Rayfran das Neves Sales and Clodoaldo Batista, to murder Sister Dorothy. Cunha testified that he was paid $24,000 by two ranchers, Regivaldo Pereira Galvao and Vitalmiro Bastos de Moura, who wanted her killed because she had accused them of illegal deforestation (Alves, 2006).

Conflicts are almost always about controlling and exploiting resources, whether rubber trees or timber or cotton or diamonds or oil, and we humans have exploited trees and each other throughout history. It might be argued that all life is exploitive, that one thing lives off another. But what makes humans different is that we exploit to extinction. If we have the capability to understand and manipulate the atom, the gene, and all the molecules of the universe, we surely might find better ways of earning a living than killing one another off to control the earth's resources and extracting those resources until there is nothing left. That, I think, is the great challenge today: to find ways of making economics work without exploiting each other and the earth's resources into extinction, to make "our human economy . . . resonate harmoniously with the larger cycles of nature" (McLaughlin 1993). For all the excellent work scientists are doing studying the ecology of the rain forest and discovering ways to conserve it (Bierregaard et al 2001), the issues of economics and morality in the Amazon seem the most difficult to solve (Slater 2003).

When we were introduced, the owner of Explorama lodges, Peter Jensen, proudly proclaimed that he was a capitalist. He was in the ecotourism business to make money. His livelihood probably represents a sustainable and nonexploitive economic endeavor. He was a smart, well-educated, well-connected, man with a likeable personality for whom capitalism works. But what about the majority of the world's people who have none of those privileges? Is an economic and political system that works only for the privileged few a denial of our ethical responsibilities? In a world full of creatures with infinite forms and multitudinous ways of living, would it be wiser to embrace as many economic forms as possible?

When we returned from birding and passed the river juncture again, the loggers were gone. They may have taken the logs to one of the lumber mills in Iquitos. I counted three mills near the docks when we first departed, and all three were busy. The thick slope of sawdust working its way down the bank into the river represented a lot of trees. Whether the trees were sold to the mills or to timber brokers, I don't know. I was never able to learn their destination.

The World Wildlife Foundation (2003) reports that illegal logging of mahogany trees is pervasive in the Pacaya Samiria preserve. A blogger named Dave (Wilderness Classroom 2006) recently reported seeing three groups of illegal loggers in the preserve cutting and sending mahogany trees downriver. But for all the concerns about illegal logging and for all the efforts at sustainability, such as the Rainforest Alliance's Smart Wood Program (Newsom and Hewitt 2005), I have never been able to specifically identify anyone doing it. I have never found the name of the companies the loggers worked for, who bought the lumber, who shipped it out of the country, what companies purchased the lumber abroad, or who the consumers were who bought the lumber. With all the layers of transactions, I even began to wonder about my own mahogany desk.

PART III

OTHER GREAT WORLD RIVERS

Mississippi River:
The Atchafalaya Basin

The day I turned fifty-nine, I was knee deep in muddy water and lost in the Atchafalaya swamp. I figured, well, at least I wasn't sixty, which had always seemed really old until I began to approach it, and then it seemed about middle age. I also thought, if I ever get out of this mess, I will buy myself a first-class ticket to Madagascar, fly in a luxury leather recliner, and see the most unique swamp in the world with a first-rate guide.

The trip had begun in Lafayette, Louisiana, as part of an Audubon Society outing to explore Cypress Island, The Nature

Conservancy preserve on Lake Martin. Avery Island, the estate of the
McIlhenny family, of Tabasco Sauce fame, was also on my itinerary.
Although both places are called bird sanctuaries, they illustrate the
contrast between a modern preserve and a cultivated garden.

Cypress Island, one of the largest wading-bird rookeries in
the United States, supports as many as 30,000 pairs of breeding
birds. Its history and beauty have been recorded in Nancy Camel's
The Nature of Things at Lake Martin (2006). The branches of bald
cypress, tupelos, and leafless buttonbush held thousands of nesting
pairs: great egrets, cattle egrets, roseate spoonbills, anhingas, little
blue herons, black crowned night herons, snowy egrets, and white
ibis. Prothonotary, yellow-throated warblers, and other warblers
darted in and out of the branches, and common moorhens meandered among the water hyacinths. We counted more than fifty
species of birds in a few hours.

Lake Martin is a typical Southern river-bottom swampland.
An eight-mile circumference trail encircles the lake and provides a
good means of exploring the terrain. In late April and May, however, the trail is closed for alligator nesting season. Among the water-
loving grasses along the lake's edge, wild flowers splashed their
colors. Bright yellow and deep purple irises were the showiest. With
five different species of beardless Louisiana *Iris*, I was unsure which
species they were, but their beauty could clearly inspire a Georgia
O'Keefe painting. Thickets of wax myrtle and hackberry comprised
the lowest vegetation, and the tall trees (bald cypress, water tupelo,
sycamore, American elm, sweet gum, green ash, sugarberry, bitter
pecan, live oak, and overcup oak) provided the high canopy. Shorter
trees (swamp cottonwood, black willow, mulberry, and swamp
dogwood) filled in the understory. They were all in their new spring
green finery.

From Rookery Road, a gravel levee road that runs along the
southern edge of Lake Martin, birds could be seen in all their breeding glory. They preened and postured, built stick nests, squatted on
eggs, or regurgitated food down the throats of fuzzy-headed chicks.

Through the lens of the spotting scope, I could hardly believe the colors: spoonbills with sunset-orange tails and rosy wings; snowy egrets with purple eye patches the color of a black eye acquired in a fight; great egrets in lacy plumes and chartreuse eyeliner; little blue herons with fluorescent blue bills; cattle egrets rouged and purpled; and anhingas with patches of sky webbing on their beaks. The lyrics of the song "In your Easter bonnet with all the frills upon it, you'll be the finest lady in the Easter parade" came to mind. How ironic that verse seemed, for snowy egrets were hunted almost to extinction because of their plumes, which were used to adorn women's bonnets.

Avery Island was a very different place. A 2,200 acre salt dome, it rises from the surrounding marsh like a green midden. Tourists pay $6.25 each to enter these "jungle gardens." While a few champion live oaks make their home on the island such as the 500-year-old Cleveland Oak, the landscape was manicured and resembled the grounds of an antebellum plantation. Open fairways with carpet grass were broken by clusters of Japanese camellias, azaleas in wilted after-bloom brown, Egyptian papyrus, and bamboo—all non-native species. The gardens were cut with canals and ponds that had little apparent aquatic life except turtles. Turtle heads poked out of the brown waters like muddy sticks. An occasional crow, a few blue jays, and a cardinal or two were the only birds around until we came to the egret pond.

On two long platforms in the stagnant pond named Bird City, hundreds of snowy egrets nested like tenement dwellers. Out in the open with no shade, they were a monoculture of white. The egrets originated from seven young birds that had been rescued and raised by Edward A. McIlhenny. The story of how young Ned saved the birds from plume hunters and established the first bird sanctuary in the U.S. is told in *The Bayous of Louisiana* (Kane 1943). I suppose the term *bird refuge* might still apply to Bird City, but how different our idea of sanctuary has become. Today, we think of a sanctuary or a refuge as a place of diversity with native species. Rather than reflecting the interests and hobbies of a wealthy white man, sanctuaries today are created by more collaborative efforts. People still visit Avery Island to see Southern gardens, but it is really a tourist attraction rather than a wildlife habitat.

Leaving Avery Island, we made our way to Whiskey Bay Road along the Atchafalaya River. A popular birders' magazine described the road as an excellent place to spot neotropical migrating warblers. My maps showed the road bordering the Atchafalaya River, the Atchafalaya National Wildlife Refuge, and the Sherburne Wildlife Management Area, ending up at Highway 190, so that was where we headed. Exiting the interstate, we found a rutted dirt road and took it. Could this really be Highway 975? I wondered. We traveled some miles, took another dirt road, then another and another. At a cypress stand that looked promising, we stopped and scanned the branches to see what birds were about. An opening into the swamp seemed like a good idea at the time and I walked for a while on what I thought was a trail. But when it petered out and I turned around to retrace my steps, the path had disappeared.

Trying to gain my direction and not watching my step, I tripped over a root, fell down a ridge, and rolled into a slough. Getting up, I kept slipping and sliding, sinking deeper into the mud, so by the time I recovered my footing, I had managed to get myself knee-deep in the muck. Crawling out of the mud, I sloshed my way through the swamp in the direction I thought was the road. I finally made it back to the car, but it was a reminder that it is easy to get disoriented in swamps. I have yet to get to Madagascar, but it is still on my wish-list of wetlands I want to see.

The Atchafalaya swamp may not be as exotic as a Madagascar swamp, but it is the largest freshwater swamp in the United States. One of five distributaries of the Mississippi River, the Atchafalaya River diverges from the Mississippi around Old Roads, Louisiana, and the basin lies south of that apex. The 2,000-square-mile basin is an elongated triangle that stretches about 150 miles to the coast and is about thirty miles wide where Interstate-10 crosses it in the middle. The view along that section of freeway might be the way a neotropical migrant would see the swamp as it zoomed through the branches of willows and cottonwoods.

John Lockwood's photographs of the basin in *Atchafalaya: America's Largest River Basin Swamp* (Lockwood 1982) are truly worth more than a thousand words, and one of the best written accounts of the region is a chapter in *The Control of Nature* (McPhee 1989). McPhee recounts the story of how the Army Corps of Engineers built a dam-valve system in the 1950s to control the Mississippi River and prevent its diversion into the Atchafalaya River. The Mississippi was about to change from its present course to the Atchafalaya with unacceptable ramifications to the cities downriver along it. The idea of New Orleans and Baton Rouge on a dry riverbed was unimaginable (although with Hurricane Katrina, New Orleans might have wished for a dry river). Today the waters are partially controlled, so New Orleans and Baton Rouge are still on a deep, navigable river, but that action has come at a price of enormous land loss.

The Atchafalaya Basin habitats run the gambit of coastal lands. The northern uplands are cleared pastures, hardwood bottomlands, and bald cypress-tupelo swamps. Freshwater marshes and lakes are numerous, and in the southernmost region, the salt marshes line the coast like a lace hemline. The coastal wetlands of the basin are one of the few places in Louisiana where land is being built. Several thousand acres per year are created from sediment deposition from the Atchafalaya River, which now carries close to a third of the Mississippi River's flow. The rest of Louisiana's coast loses nearly 20,000 acres a year (Tidwell 2003). The building of levees along the Mississippi has directed the river's flow so that the rich sediment from the heartland upstream drops out and falls over the continental shelf in the Gulf of Mexico. The sediment is lost for land building. That is the price we pay for flood control, such as it is. Access canals all over the Louisiana coast have also eroded the wetlands. That is the price (one of many) we pay for oil.

A canal, of course, is an expedient thing, with its straight and focused direction, so unlike a river. A river meanders. It curves and bends, it loops back on itself and drifts along. It changes direction. It makes oxbow lakes where it abandons an old channel and forms a new one. It floods and creates lakes, ponds, and streams. It builds land when left to its own devices. It might be easy to think of canals and rivers like certain types of people. Some people move through life like a canal: straight, directional, focused. Others move like a river: meandering about and exploring in a roundabout way. Both river and canal end up at the same sea, but a river's way of getting there is so much more interesting. In truth, at least in broad biological truth, a river's way is far more productive.

Thirty-five years ago, I traveled the basin from its upper end to the coast, from Henderson Swamp to Marsh Island. The field trip was part of a wetlands ecology course. It was late March, and in the words of Longfellow, it was

> the forest primeval . . . where plumelike
> Cotton-trees nodded their shadowy crests . . .
> [and] towering and tenebrous boughs of the cypress
> Met in a dusky arch and trailing moss in mid-air
> Waved like banners . . .
> (Longfellow 1922, 1, 62, 63)

The red swamp maples hung with fringy auburn fairy wings; seeds of these maples are one of the few that form and get dispersed in spring. Shawls of Spanish moss draped the spring-green tupelos, and yellow catkins dangled from the oaks. The open fields along the river sprouted columns. One bank was covered with the mud chimneys of crawfish; another was full of bull thistles. Across the river banks, cypress knees rose like brown stalagmites.

The river was in bloom. Yellow tops or butterweed (*Senecio glabellus*) created a Monet waterscape, and a marsh of nothing but bull tongue (*Sagittaria lancifolia*) undulated like green waves. In the backwaters, those Attila the Hun invasives, water hyacinth (*Eichhornia crassipes*) and hydrilla (*Hydrilla verticillata*), formed tangled mats and clogged the channels. Mud-loving fish like sacalait, gizzard shad, goggle-eyes, gar, bass, and catfish drifted in the sluggish dark waters beneath the green floaters.

The second leg of the field trip took us to Marsh Island. At the southernmost end of the basin and the river's restitution, Marsh Island lay a little west of the mouth of the river. A nearly treeless mound of about 70,000 acres, it is all refuge and state-managed land. Weirs and sill dams help retain freshwater and promote wigeon grass (*Ruppia maritima*) for wintering blue and snow geese. The island also supports a healthy population of other water fowl as well.

The island has an interesting history. It was purchased by Margaret Olivia Slocum Sage, the widow of millionaire Russell Sage, who then gave it to the state of Louisiana for a bird sanctuary. Russell Sage, a notorious railroad baron and Wall Street banker at the turn of the twentieth century (Sarnoff 1965), left his fortune to his widow. Believing that women were the "moral superiors of men," Margaret Sage established a philanthropic foundation and a college with his money, and also purchased Marsh Island and gave the island to the state. For nearly a hundred years the island has been a wildlife sanctuary courtesy of Mrs. Sage. Even though the refuge is called the Russell Sage National Wildlife Refuge, it more rightly could have been called the Margaret and Russell Sage Refuge. Regardless of its name, I find a certain sense of justice knowing that a robber baron's money went to buy public land for a bird sanctuary. Praise be, the wisdom of some wives!

Crossing Vermillion Bay from Cypremort Point, a strong northwesterly wind churned the waters, making it a choppy ride even in a Boston whaler, and I remember getting a little queasy. We docked at the state's Wildlife and Fisheries Station, picked up the resident biologist, and proceeded out into the marsh. The marsh was a textbook example of the many different coastal grasses that grew along the Gulf Coast. The biologist rattled off the names of the plants like an auctioneer: salt-tolerant oyster grass or cord grass (*Spartina alterniflora*), salt grass (*Distichlis spicata*), brackish water wire grass or salt meadow cord grass (*S. patens*), big cord grass (*S. cynosuroides*), phragmites (*Phragmites australis*), needle rush (*Juncus roemerianus*), and sedges (*Scirpus* species). His litany of grasses, spoken with the same familiarity someone might use in calling their children, was a song of true affection.

The half-dozen students on the field trip were marine science and ecology majors. One was a nerdy undergraduate named Cecil or Malcolm or Horace or some such name that has long slipped from my memory. We were about to step out of the boat onto a spoil bank when we spotted a huge water moccasin coiled and ready to strike. Its grey body was as thick as a man's arm and it was very annoyed at being disturbed. Some of the boys tried to get the snake to strike; some wanted to kill it. They all made like they would push it into the boat when they saw how freaked I was at the sight of that monster cottonmouth, but the nerdy kid used a boat paddle, gently moving the animal off into the grass and it slithered away. He has remained in my memory as the boy who gently chased away the snake. I imagine he may be a River Keeper somewhere today. Who knows, maybe he's a snake handler in some religious group. In Louisiana, anything is likely.

The Mississippi River has played such an important role in American culture. East and West were historically marked by its presence and still are today. America's great waterways, crucial civil war battles, the post-World War II industries, and many of our great American stories including those of Mark Twain (1883), have taken place along the Mississippi River. There is something about the river that is so quintessentially American. Like our country, its lower deltas are relatively young lands in geological terms—about 50,000 years old. The land is vibrant, still forming, still evolving, and changing. That is the nature of deltas.

Because deltas are fertile, most have been turned into cultivated farmlands. The fertile Atchafalaya Basin supports soybeans and sugar cane, but its wetlands have remained relatively undeveloped. Unlike the Everglades, whose wetlands have been drained and turned into crop fields and housing developments, the

Atchafalaya Basin still remains in part a natural place. When we examine the history of what people have extracted from it and how it has been exploited, however, it is a wonder anything natural has remained.

When we look at the human impact on the Atchfalaya Basin, the place that probably best illustrates those activities is Morgan City. The city at the end of the basin is famous for its Louisiana Shrimp and Petroleum Festival, held every Labor Day. Is there any other city in the world that would celebrate two such incongruous industries at the same time? How did this come to be? A look at the city's Web site (Morgan City 2005) gives a brief history.

The Chitimache Indians were the aboriginal peoples of the Atchafalaya Basin, but by the 1700s, when French settlers moved into the region, disease and war had decimated their population. By the late 1800s there were only six Native American families left. In the mid-1700s, the French Acadians, or Cajuns, settled in the area. Their story is romanticized in Longfellow's well-known poem, "Evangeline." In the mid-1800s, Morgan City, then called Brashear City, became a sugar plantation owned by Walter Brashear, a rich Kentucky planter and legislator. Imported slaves provided the backbone of that economy. After the Civil War—referred to on the Web site as "a rich man's war and a poor man's fight"—and during reconstruction, another system replaced the plantation economy. Charles Morgan, a railroad baron and steamship magnate, dredged the Atchafalaya Bay Channel and opened the city as a port. It then became a trade center for cypress timber, fur, seafood, and boat building. In the 1930s, Morgan City was known as the jumbo shrimp capital of the world, and by the 1940s, with the discovery of oil, another boom economy began.

Today the ancient cypress trees are gone, the enormous populations of fur-bearing animals are gone, and a logical question might be, Will the seafood and oil industries be the next to disappear? Canals, hydrological changes, and other oil-related activities may destroy the renewable seafood industry. The nonrenewable resource, oil, will certainly give out some day. Maybe other industries will replace those, and maybe the land will sustain them. Unlike the Nile delta, which has been impacted by humans for thousands of years, the Atchafalaya delta has experienced only a few hundred years of intense human intervention. The land has made a lot of men wealthy, but there may come a time when it can no longer sustain that use.

Leaving the Atchafalaya Basin and returning home from our birding trip, the last stop was at the welcome center along the interstate. Built in pseudo-plantation style, the large new building was surrounded by an enormous parking lot. The wetlands had been filled in with clay, and the building was erected on the spot where a small cypress grove had once stood. There was no one in the welcome center, but the brochures that lined the walls were geared for tourists and home buyers. That real estate and tourism have become the latest economic boom troubled me. I am not as concerned about visitors, but I shuddered to think of the basin being turned into a subdivision. As we left the rest stop, I thought, maybe another wealthy widow will come along, buy the whole basin, and give it to The Nature Conservancy for a bird sanctuary. Who knows, miracles do happen, even in Louisiana.

Danube River:
The Tizsa and the Hortobagy

O ur group of birders was traveling to the Hortobagy, a preserve on the Great Hungarian Plain, to explore the steppe lands known as the puszta. The motorway from Budapest to the Hortobagy stretches about a hundred miles through eastern Hungary, a country the size of Indiana. It was late October and the fields were brown. Melons lay in some fields while dry corn stalks and sunflowers made up the fall texture of the others. The black sunflower heads, bent over, all facing one direction, resembled a mosque full of devotees in prayer.

The Hortobagy Nemzeti Park, Hungary's first national park, lies in the middle of the country in the Tizsa River floodplain. A tributary of the Danube, the Tizsa River begins in the northern arch of the horseshoe-shaped Carpathian mountain range in the Ukraine, flows southward through Hungary and joins the Danube near Belgrade. One of the best descriptions of the region was written in 1933 by Patrick Leigh Fermor in *Between the Woods and the Water: On Foot to Constantinople from the Hook of Holland: The Middle Danube to the Iron Gates* (Fermor 2005). At the age of nineteen, Fermor traveled across Europe, and his observations seem as relevant today as they were seventy years ago.

Our exploration of the Hortobagy began with a trek through the soggy rape fields in search of great bustards. The ground was rough and full of puddles. Frogs were everywhere. Green marsh frogs, tree frogs, and spade-foot toads hopped up and out in all directions as we trekked through the wet fields. Clouded sulfur butterflies flitted among the grasses, and the outline of a side-arm well far off in the distance was a great upside-down Y of wooden poles. A couple of roe deer darted off, disturbed by our presence. The tri-colored landscape was white from the millions of lacy spider webs that had collected moisture overnight, green from the wet fields, and grey from a sky the texture of heavy gauze.

The Hortobagy Nemzeti Park, established in the 1970s and now a UNESCO World Heritage site and Biosphere Reserve (World Heritage 2005), is about twice the size of the Bosque del Apache National Wildlife Refuge in New Mexico, or about 130,000 acres. There are also four conservation areas around it that extend its size. Its grassland ecosystem resembles the great Asian steppe. The soil is salty—it's called solonetz soils—and alkaline due to the drainage pattern and evaporation of ancient rivers. But the park is more than a wildlife refuge. It is a multifunctional preserve serving as mostly pesticide-free pasture and agricultural land (40 percent), fish farms (20 percent), and wildlife refuge (40 percent).

After a successful search for bustards, we headed to our lodge for the evening. Herds of grey Hungarian cattle grazed in the fields, and long spiral-horned sheep called racka, indigenous to the region, were being rounded up by shepherds and their great white dogs. The dogs, called komondors, were large shaggy-coated beasts that resembled long-legged sheep, and their evening round up seemed to be in full swing as they encircled the sheep into an ever-shrinking pack. Dinner in the hunting lodge was goulash served with enormous slices of freshly baked bread. Thick with beef chunks, potatoes, carrots, beans, onions, and small noodles that tasted like dumplings, with just the slightest hint of paprika, it was an entirely satisfying meal after the damp chilly day on the grasslands.

Morning began with a misty walk around a nearby pond. A little bittern moved slowly through the reed beds, barely visible in its camouflage stripes. Overhead, the common cranes' honking sounded like a wedding parade as they moved from roosting sites to feeding grounds. The road leaving the lodge was lined with sugar beets, freshly harvested and still covered with dirt. They looked like stones piled ten feet high to form a wall. The small towns, villages, and hamlets along the route were full of neat whitewashed houses, some painted soft colors, all with red tile roofs. Stout and sturdy homes enclosed by fences, they seemed self-contained and self-sufficient microcosms. The front yards had vegetable and flower gardens; the back yards, chickens and barns. On the top of tele-phone poles, perched like big laundry baskets, were stork nests. Some had grass growing from them. They would be repaired and reused when the birds returned in spring.

A long trek through the spongy puszta in search of dotterels gave me the chance to observe the great diversity of grassland plants on the treeless plain. What a surprise to see plants that reminded me of my coastal North American home. There were salt-tolerant plants like needle grass (*Stipa capillata*) and bunchgrass or fescue (several species of *Fescue, F. pseudovina* most common), and grasses like blue stem (*Andropogon ischaemum*) and crested wheat grass (*Agropyron cristatum*). Other plants like sea lavender or marsh rose-mary (*Limonium gmelini*), yarrow (*Achillea asplenifolia*), clover (*Tri-folium* species), plantains (*Plantago* species), meadow rue (*Thalictrum*

minus), wormwood sage (*Artemisia* species), yellowstar thistle
(*Phlomis tuberosa*), purple field aster (*Aster tripolium* and other species of asters), and so many more filled in the meadow's mosaic. In
some spots where water stood, the groundcover was a crusty lichen
similar to tundra vegetation. The grassland was an enormous canvas of plants beneath an open sky so wide it seemed to stretch
forever. Far off in the distance, the barns and buildings of Angel
House Buska were tiny knobs on the horizon. With the clanging
bells and the ever-changing amoeboid herd of grazing sheep, we
walked for hours across the great prairie in search of dotterels.
Only when everyone had a good long look at the birds and our feet
were totally soaked did we move on.

It was evening when we arrived at Pali Halom Lapos, where
we watched the cranes fly in for the night. Our perch was a small
mound called a tumuli. Tumulis are ancient burial mounds scattered throughout the region. Left by Neolithic people who once
roamed the grasslands, this one had not yet been excavated, and it
provided us with an elevated overlook of the plains. From our
perch we could see cowboys (called csikos) rounding up cattle and
driving them to corals. Their beautiful horses, manes and tails flying
in the wind, moved with such grace and wildness. It was easy to
understand why the region was known for its horses. One line of
the famous Lipizzaner stallions makes their home in nearby Eger.
Off in the distance our guide pointed towards the horizon and said
there was an ancient salt road that ran along the ridge, a road that
connected Western Europe to the mines in Transylvania. Clearly,
this land had been inhabited by humans for thousands of years and
was an important trade route between East and West.

As the sun sank lower on the horizon, strings of cranes in the
thousands drifted down against the evening sky, and their honking
filled the air. A mist began to blur the skyline, and the cranes' sky
writing grew fainter. Their flight scribbles continued but faded as
the entire plain turned dark. Well into the night, their calls trum-
peted over the grassland. About ten thousand common cranes were

in the preserve, but as many as thirty thousand would arrive with oncoming winter. The cranes use the Hortobagy as a stopover and staging area for their winter migration. In spring the cranes migrate to their breeding grounds in Sweden and Finland; in winter they move south into North Africa and the Middle East.

The fish ponds were our next destination. They varied in size from small quarter-acre ponds to large lakes connected by ditches, canals, and pumping stations. The levees surrounding the ponds were covered with typical wetland plants such as willows (*Salix* and *Salicetum* species) and reeds (*Phragmites* species and *Scirpus* species). The landscape seemed so familiar that I felt like I was home in the wetlands of the Mississippi flood plains. Plants in the Hydrocharitaceae family (water weeds like *Elodea* species) were abundant. Wild rose vines (*Rosa* species) covered with red fruits known as rose hips grew like tangled hedgerows, and the berries of hawthorn (*Crataegus* species) reminded me of southern mayhaws. They were vivid holiday red and lusciously plump. Shrubs such as western hackberry (*Celtis reticulata*) and others I did not recognize grew along the ponds' edges. There was thorn apple (*Datura* species), aptly named because of its prickly thistle-like leaves and fruit, marshmallow (*Althaea* species) with its white crepe-paper ruffled flower, and an occasional white campion (*Silene alba*).

Clusters of a tree known as the tree of heaven (*Ailanthus altissima*) grew near the embankments, and I remembered this sumac-like tree as an invasive. Native to China, it is an exceptionally fast grower, and with its willowy shape and spindly fern-like leaves, it became popular as an ornamental shade tree. Because it grows so quickly, it has became an invasive, out-competing native species and becoming a nuisance in many regions of the United States, even in Brooklyn, where it inspired *A Tree Grows in Brooklyn*.

The fish ponds were used for the aquaculture of carp, and I noticed a dead one floating near a partially submerged boat by the shore. It looked like the body of a drowned child, it was so huge. To grow such enormous fish requires nutrients, and the ponds were fertilized with cow dung from local farms. As we walked beside the ponds, several trucks passed carrying loads of dung. They dumped the dung in high steamy black piles along the banks. From the pungent piles sprang beautiful long-stemmed white mushrooms capped with black stars. No doubt about it, cow manure seemed to make everything grow. The thought occurred to me that dung had always been used in so many practical ways, from fertilizing and growing crops to making fire, but its wondrous life-giving properties did not diminish its unpleasant odor. I was happy when we left the ponds, glad to be rid of the smell, yet equally thankful for the richness of our bird list due in part to the nutrient-rich manure.

Lunch along the Tizsa River at a riverside café was fish soup. Made with catfish in heavy cream, the local patrons seemed to love it. A buxom waitress kept bringing tureen after tureen full of thick white soup and platters piled high with bread to the table. I tried not to be too obvious in my lack of interest in the heavy fishy soup, but I still had the odoriferous dung piles imprinted in my smell center, and fish in cream has never been one of my favorite foods. After lunch, we walked to the river and watched it flow on its slow way to the Danube. I recalled an incident referred to in European newspapers as the "Chernobyl of Hungary" (Kiss 2000). In January 2000, a massive cyanide spill occurred on the Tizsa River. The tailings dam of an Australian-Romanian mining company, Esmeralda Exploration, broke and released 3.5 million cubic feet of mining waste waters, containing cyanide and heavy metals, into the river. Massive fish kills of over a thousand tons and bird deaths were recorded along a fifty-mile stretch of the river near the spill. From 30 to 40 percent of all biological life was destroyed along the river, although zoo- and phyto-plankton seem to have recovered quickly after the cyanide plume passed. Five years later, biologists mostly agree that the ecology of the river has not recovered, and it may take decades for it to do so. After the spill, numerous task forces were established and efforts were made to enact legislation to prevent future spills and provide restoration. Some of the cross-national boundary efforts and European Union actions included legislation to better regulate mining wastes and prevent design flaws. An Environmental Liability agreement was also enacted that allowed affected individuals the right to compensation, in other words, to sue. The long-term effects of the toxic waste may not be known for many years, nor will the results of efforts to restore the river be evident very soon, yet there was deep concern and the political will among the citizens to prevent such incidents from happening again.

Afternoon involved another trek through the grasslands, but in this section of the preserve, it was wooded steppe. Deciduous trees such as the indigenous tartar maples (*Acer tataricum*), species of oak, common elm, ash, elder, and black locust were almost leafless. We walked past a thin grove of poplar trees (*Populus albus* and *P. canescens*) where thousands of rooks had gathered in the bare branches like lumps of coal. They created such a racket with their harsh caws. In the crow family, the corvids are colonial nesters, and the term *rookery* comes from their nesting behavior. Thought to be the most social and intelligent of the birds (Hess 2005), rooks certainly seemed the most vocal. Eventually, the flock flew off and colonized a small pine forest in the distance, taking their raucous roar with them.

Our final observation of the day was near the town of Balmazujvaros in the northeastern edge of the preserve. It was a roosting site for long-eared owls. Initially, we could see only a few owls, maybe six or seven, in the dense tree branches, but as we continued to look, the count went up to ten, then fifteen, eighteen, and finally twenty-one. As many as 300 owls have been reported to roost in the trees later in the season as the cold settles in. Located in a fairly heavily populated suburb, the trees seem an unlikely place for owls, and I wondered if people disturbed them. Since the trees and the small plots around them had been purchased by the state, they were protected habitats. But I wondered how effective such protection really was. As we scanned the branches trying to locate the well-hidden owls, some of the neighbors came out from their town houses with binoculars and smiles. From their gestures, it was obvious they relished the owls as much as we did.

As we left the owl roost and returned to our lodge for our final evening, I marveled at this wondrous preserve. The Hortobagy represents the best experiment in sound ecology. It is an economically sustainable, multifaceted preserve blending pasturing, agriculture, and aquaculture; it maintains the region's ethnicity and cultural heritage; and it provides a habitat for many threatened bird species and wildlife. What more could you ask of any land? The citizens of Hungary obviously care about their environment, and they have the political will to keep it healthy. That Hungary, the European Union, and their international partners have established and maintained such a remarkable reserve reflects a wise and visionary citizenry. Nearly 150 years ago, Johann Strauss wrote the lovely composition, "The Blue Danube," in honor of the river. I wondered if he had any idea that one of the river's tributaries would someday hold a place with as much beauty, harmony, and balance as his waltz.

Yangtze River: River of Change

BEIJING

P aul Theroux's description of Peking (Beijing) in *Riding the Iron Rooster: By Train Through China* (1988) as a city on the rise should have prepared me for a modern city. Instead of streets full of slow-moving bikes and modest buildings, I was surprised to see a landscape of sky scrapers, concrete high-rises, cars, buses, smog, and congestion; in short, a metropolis comparable to any large city in the world. Six freeways circled the city as beltways, each one located a bit further from the center in concentric rings. It can take six hours to go from one end of the city to the other, Lily, our guide, said. And she laughingly remarked that, "In Beijing, it is always rush hour."

From the airport, we sped along the smoggy freeway without serious delays and were in the heart of the city in less than an hour. It was winter and everything was grey. The dendrite pattern of bare poplars made a graphite line drawing on a slate-colored background. The pattern was broken by magpie nests. Great tangles of sticks lay in the crotches of some of the branches. Occasionally, a bird the size of a large crow entered or emerged from its nest or glided from one tree to another. At our hotel, these black-billed magpies soared from the ledges of a high-rise apartment to the shopping mall to the hotel and back as if the surrogate canyon cliffs were a perfect home. Birds of joy and luck in Chinese mythology, they seemed to be a good sign to begin our travels in the country known as the Middle Kingdom.

For the next few days we were typical tourists in the smoggy capital. All of China's cities are infamous for their coal-burning smog, and Beijing, surrounded by its huge industrial zone, offered no exception. The smog from the factories, especially the Shougang steel mills, settled over the city like a frosty bowl, and we never experienced a blue sky day. Supposedly, the steel giant intends to relocate one of its plants and suspend production during the 2008 Olympics (Xinhua News Agency 2005) but such temporary measures represent no real solution. China's enormous economic modernization in less than twenty years has not come without the high price of environmental degradation. One of the best accounts of China's environmental crisis is Elizabeth Economy's *The River Runs Black: The Environmental Challenge to China's Future* (2004).

Each morning Lily, clutching an English copy of *1421: The Year China Discovered America* (Menzies 2002), met us with a smile and took us to all the places a tourist would want to see: Tiananmen Square, the Forbidden City, the Summer Palace, and the Great Wall. After a whirlwind of tours, we flew from Beijing to Yichang to begin our journey up the Yangtze.

ON THE RIVER

Two young women, a driver and guide, met us at the Yichang airport and transported us to the city. Their minivan was a standard shift, and when the young driver shifted gears, the van bounced along like an old school bus. Sliding around on the slippery vinyl seats, I was reminded of the cold hard seats I endured as an elementary school kid. The landscape rolled by like a green mural. Even in winter the hills were covered in orange trees and vegetable fields. The many shades of green made a multi-textured tapestry, and the terraced fields swirled up the slopes like green staircases. Too soon, the countryside ended and we were in another smoggy, congested city. The smog hung heavily over every street and I wondered if the people of China, over 45 percent of whom now live in cities, missed the winter sun. When our guide asked if we would like to see the museum displaying relics from the Shang Dynasty, we said sure. We had plenty of time before boarding the boat, and it seemed like an interesting side trip.

The drab concrete building, sooty from the coal burning, was not what I had expected. A trim middle-aged woman in a grey-green uniform welcomed us at the door and took us through the museum at a breakneck pace. Speaking in rapid, programmed English, she practically sprinted through the exhibits. The displays went by so quickly, they seemed like a fast-forwarded film. She explained that the artifacts excavated from nearby sites would be lost with the completion of the Three Gorges Dam and the flooding of the land. Almost before we had started, we were in the "gift shop," where she began a heavy sales pitch. Bronze statues, jade fetishes, and other "relics" thousands of years old were very reasonably priced, she exclaimed. As the director of the museum, she would be able to give us a substantial discount. The museum had to sell the artifacts (complete with documents authenticating their age and legality) to sustain itself.

I was beginning to feel like a character in the mystery, *Dragon Bones* (See 2002), which tells the story of the shady trading in archaeological artifacts and especially the so-called prized "oracle bones." I was annoyed, since I thought we were visiting a museum and not a marketplace for relics. Ancient artifacts should not be sold and taken out of the country. While I know people in great financial need do not always have the luxury of such opinions, I wanted no part of such transactions. All the antiquities—a bronze hawk, a jade turtle, and ceramic vases—looked like the mass-produced junk in any souvenir shop. But I felt badly for the woman when we left without buying. She looked so dismayed when we refused to purchase anything, as if our actions would result in the immediate closing of the museum and the personal loss of her job. In Beijing, we had been approached often with offers of items from tea to trinkets, rugs to watches. The whole experience was beginning to reveal a China adopting the worst elements of capitalism and accepting the attitude that everything was a commodity and a marketable product. I was aware that China was evolving into an economy referred to as "capitalist communism" and that the old saying by Deng Xiaoping, "It doesn't matter if the cat is white or black, as long as it catches mice," was being put into practice, but this frenetic commercialism was annoying on a personal basis and troubling on a broader level.

Our boat, *The Victoria Empress*, was docked behind a hotel. It seemed disconcerting to have our luggage carried down the steep river bank by two tiny women the size of children. They probably weighed collectively less than a hundred pounds, but they lifted our suitcases onto their heads and trotted downhill as if the bags were pillows. Our cabin, one of about seventy, was on the second level and as cold as an ice box. We huddled under the bedcovers until dinner, unable to figure out how the heating system worked. The dials had no labels or markings of any kind. We requested that the staff turn on the heat, but whether they did not understand our request, the ship's heat was not on yet, or we just did not turn the right buttons, the result was a freezing cabin. It was so cold we slept in our clothes—coats, shoes and all—and the next morning we awoke stiff and in much need of hot tea. As the boat departed from Yichang, one of our tablemates, an engineer, explained how to work the heating system, and we soon had a toasty cabin, a must for a comfortable journey on the Yangtze in winter.

The next morning's excursion to the dam site at Sandouping began with the distribution of umbrellas. It was a dreary day, and rain muted the kind of scenery only an engineer would love. The landscape looked like the biggest strip mine in the world. High cranes, scaffolding, a concrete coffer dam, and dug-up earth were all part of an engineering feat of monumental dimensions. Mark Levine's description of the site was accurate, "Everywhere I look the skin of the earth has been removed and what remains is gristle and pale fractured land . . ." and "this new great wall, a 26-million-ton slab of concrete . . . will rise as high as a sixty-story building above the riverbed and span a mile and a quarter from shore to shore" (Levine 2002, 199). As an ecologist, I know dams always mean significant disturbances to the natural environment, disruption of the natural hydrology of a river, and the destruction of wildlife. But with a population of 1.3 billion people who need more energy and more control of the river's devastating floods, China's decision to build such a dam seemed inevitable. There was probably more at issue than electricity and flood control; the success of this project might be equally important to China as a political symbol. Having been colonized and humiliated by the European colonial powers in the nineteenth century and the Japanese in the twentieth century, the largest dam in the world might reflect a great national achievement. As Simon Winchester claims in *The River at the Center of the World* (1996), it may be China's "New Great Wall."

Having spent the night in the cold, I was far more receptive to the argument that the dam would provide much needed kilowatts to a population that generally does not have central heat in their homes. But the displacement of the 1.2 million people who lived along the river was unimaginable. The flooding of 1,400 villages and the loss of millions of fertile acres was mind-boggling. The stagnation of the reservoir, the loss of fertile sediment downstream, and the subsequent land loss were irrefutable arguments against the dam. A number of Chinese environmentalists (most notably, the investigative journalist, Dai Qing) have criticized the dam at great personal risk. Provoking the wrath of a government that does not tolerate dissent is always dangerous. After Dai Qing published *The River Dragon Has Come: The Three Gorges Dam and the Fate of China's Future* (Dai Qing 1997), which presented the views of forty top Chinese scientists, engineers, and hydrologists who opposed the dam, she was arrested and sentence to prison. She spent ten months in Beijing's notorious political prison, including six months in solitary confinement. All of the arguments and all of the dissent have been to no avail. The dam continues to be constructed and the waters have risen to about two-thirds of their destined mark. In 2009 when the dam is completed, the final gates will be closed and the waters will rise to the final 177-meter mark.

The shuttle bus groaned up the hill and dropped us off at a park overlooking the dam site. The park provided a bird's-eye view of the dam and lock system. Inside the pavilion, a model of the massive construction showed the resulting lake in miniature. While most of our fellow travelers circled the exhibit, intrigued by the model of the controlled river, we slipped off to explore the gardens. A flock of light-vented bulbuls flew among the bare branches. Perched on the rain-drenched branches, they looked as damp and drab as the weather. Since they were our first sighting of flycatchers, we were excited, but we would soon discover they were as common as sparrows and could be seen everywhere along the river.

It took several hours to navigate the locks. Ships were lined up like boxcars: freighters carrying huge cargo trucks packed to the gills, coal barges, grain ships, and passenger ships all squeezed into the high concrete-walled canal. It was a massive movement of goods. With only a few feet of space between boats, it felt claustrophobic.

It was easy to understand why the Yangtze River was known as China's major transportation artery. Gone were the junks, sampans, wupans, and small fishing boats. What moves through the river now are freighters, tankers, and cargo ships as modern and industrial-size as the high-rise cities of steel and concrete that girdled this liquid highway.

As we made our way upriver, the shoreline changed from heavy industry to an agricultural landscape. Of the current natural histories and portraits of the Yangtze River, probably the best are *The River at the Center of the World* (Winchester 1996), *Before the Deluge: The Vanishing World of the Yangtze's Three Gorges* (Chetham 2002), and *Yangtze Remembered: The River Beneath the Lake* (Bulter 2004). These writings describe a river 3,965 miles long with an immense watershed. Below the Gorges and the dam, the river flows eastward through a thousand miles of lowlands and empties into the East China Sea near Shanghai. The flat eastern agricultural lands, fed by the nutrients of the river, are what most Americans think of as China, with its rice paddies and farmlands. Known as the land of fish and rice, the region represents about a third of the river's course. Two-thirds, or almost three thousand miles, of the river traverse western China and the Tibetan plateau. It is this vast, unfamiliar region that Simon Winchester (1996) describes in his travelogue. Our river trip would be along only a tiny section of the Yangtze, a mere three hundred miles from Yichang to Chongqing, where the Yangtze cuts through the Wushan Mountains and forms the Three Gorges. This region has been described by both Chetham (2002) and Bulter (2004).

Although the terrain is mountainous, it seemed that every possible inch of land along the river was under cultivation. The terraced green fields of winter wheat, cabbage, potatoes, peanuts, mustard, turnips, and other vegetables sloped up to the very top of the mountains. Where no crops grew, frost maples filled in the spaces. Their lacy limbs turned the mountainside into a cross-stitch of auburn. Trees too far away to identify filled in the sampler with yellow, dun, and crimson leaves. The bare trees may have been peach or apricot, as the region is known for these fruits. Through my binoculars I could make out the dark green foliage of the citrus trees loaded with oranges. Varieties such as mandarin sweet, peach leaf, goose egg, navel, summer, and blood red oranges speckled the slopes like party balloons. Spruce and firs added another shade of dark evergreen to the hillside tapestry.

As we moved upriver and the fields drifted by, I thought about how long the land here had been cultivated. Archaeologists believe that this region and the one to the north around the Yellow River were where agriculture first began. To imagine a land that sustained people for thousands and thousands of years, and to think of a people whose ancestors went back to the very beginning of farming, was to contemplate the intimate connection between humans and the land around rivers. How appropriate that many of our creation stories involve humans made from clay and river water—we are so much a part of the earth and its rivers. That essential connection seemed so clear here, even in the dead of winter. From the soil, river water, and people's labor came sustenance. When those who worked the earth departed from the living, they returned to the earth, their bodies becoming part of the land that sustains the next generation. It seemed so simple. The people of China honored their ancestors, had such respect for their land, and revered their river. They were part of an eternal cycle. And perhaps they still understood those crucial connections. Or perhaps they would abandon the notion of the good earth for a worldview where technology becomes the supreme ruling force.

In late afternoon, our boat entered the first gorge, Xiling Gorge. The river narrowed, and it began to look like a Scandinavian fjord. The waters turned charcoal and the banks steepened. The rock walls rose thousands of feet into the slate sky. Mist muted the riverscape, and as the river constricted into a tight hallway of stone, the surroundings became many textures of winter greys.

Darkness came early in mid-December, and by six that evening the shoreline had faded except for the occasional string of lights. When evening closed down the landscape, we turned our attention to activities onboard. The ship's staff provided our first evening's entertainment, a fashion show. They slipped into colorful costumes and paraded across the stage as Qing emperors and empresses. From drab uniformed housekeepers and servers, these young men and women, who had been almost invisible during the day, became creatures of beauty. What lovely transformations! They strutted and sashayed into the stage lights in their bright clothes. When the evening fashions ended and the taped music grew quiet, I ambled off to bed with red and gold silk dragons of ancient dynasties shimmering in my mind.

The morning began at six with Tai Chi lessons. We gathered in the ship's lounge, a sleepy regiment, and tried to imitate our instructor, Dr. Wu, the ship's doctor. Dr. Wu moved like a willow branch in a gentle breeze; we moved more like rotten limbs. After the Tai Chi lesson and breakfast, the ship entered the second gorge, Wu (the Witches) Gorge, and I scooted up to the top deck to get a better view. Rugged, jagged peaks rose high into the grey sky. The clouds opened momentarily to reveal a few patches of blue, but they did not last long. It seemed grey clouds were determined to be our companions on this voyage. Twelve peaks in all, the pinnacles of the Wu Mountains went by various names which I could not even begin to pronounce much less remember. One called "The Goddess" looked a little like a woman in a long gown walking across the sky. Paul Theroux in *Sailing Through China* (1983) claimed that every rock along the gorges had a name. I think he was being facetious, but maybe not.

Around mid-morning the ship arrived at Wushan, a modest-sized river town of about 30,000 people. We docked and transferred to a smaller boat for an excursion up the Daning River. Bundled in our coats and scarves, we climbed aboard and roared off to the Three Lesser Gorges. The boat glided beneath a high bridge, rounded a bend, and the city disappeared. What took its place was a lush green landscape. The waters turned the color of jade, and the banks became walls of leafy malachite. Even though it was near freezing, the vegetation looked tropical. Dense mosses and grasses grew on the high banks, and banana trees, thickets of bamboo, and lush shrubs sprouted in abundance.

In the first of the Lesser Gorges, the Dragon Gate Gorge, the cliffs of sandstone and underlying limestone held hanging gardens. On the rock face, unfamiliar grasses, mosses, and plants resembling wild scallions and garlic sprouted from the stone. I recalled that a mushroom called lingzhi (*Ganoderma lucidum*) grew wild in these hills. The red shelf fungus possesses many phytochemicals ranging from cholesterol-lowering substances to anticancer properties. According to several biochemical studies, the bracket fungus contains anti-inflammation compounds (Stavinoha and Satsangi 2005) and immunomodulators known as cyctokines (Halicka et al 1997). With my binoculars, I scanned the hardwoods along the banks in hopes of spotting the red lacquered fungus on the base of the trunks but saw nothing even remotely similar to it.

The region is well known for its medicinal plants, and over a hundred different species used in traditional Chinese medicine are harvested from the area. Plants like cliff cabbage (*Bergenia purpurascens*), fritillary (not the butterfly but the bulb of *Fritillaria pallidiflora*), dangshen (the tuber of the vine-like *Radix codonopsis*), Chinese goldthread (the rhizome of *Coptis chinensis*, a plant in the buttercup family containing the antimicrobial alkaloid, berberine), species of pharmacological magnolias such as *Magnolia liliflora* (the lily magnolia), *M. obovata* (a bigleaf species), and *M. officinalis* (the houpu tree), plus many others are native to the region. So popular have some of the indigenous magnolia species become that they are threatened by the over-harvesting of buds and bark for use in therapeutic extracts. In addition to the medicinal plants, rare ones like the dove tree (*Davidia involucrata*), a species in the tupelo family, are found here and nowhere else in the world. In spring, large white bracts beside the flowers give it the appearance of doves in flight. In the United Kingdom, the tree has been called the pocket handkerchief tree, as the bracts hang down like white hankies.

High caves cratered the Misty Gorge, the second Lesser Gorge. Our river guide, Mr. He, pointed out a coffin in one of the high caves. He explained that very little is known about the ancient Ba people who lived in the river lands over five thousand years ago, but they buried their dead in wooden coffins and suspended them in the caves along the river. Sometimes, the coffins looked like boats or canoes, but this one had a simple box shape. The particular cave he pointed out was one of more than seventy along the river where coffins had been discovered. It occurred to me that these early river people might have considered the river a highway to the afterlife. They may have made the connections that their lives were like the seasons, winter as death, spring as resurrection, and with the rising of the river, new life. Could they have imagined the Daning River like, the River Styx or the River Nile, carrying them to an afterlife?

Along the rocky slopes, a chiseled path was visible above water level. It appeared and disappeared in an undulating pattern of submersion and immersion. The remnant trails were the trackers' tow paths. John Hersey's novel, *A Single Pebble* (1956), gives a glimpse into the lives of men who for thousands of years pulled boats up and down the river along these stone ledges. The narrator of his story, a young American engineer who comes to build a dam on the river, travels up the Yangtze on a junk pulled by fifty trackers. The head tracker, a fellow simply called Old Pebble, is a man of great strength and purity and represents the engineer's antithesis. The tension between the two men might be interpreted as that between the east and west, the traditional and the modern, superstition and science. Initially, I thought it was a very Chinese story, but it ends not in the harmony of existence between these two worlds, but in the destruction of the old tracker as he slips and falls to his death in the treacherous waters.

Trackers disappeared in the 1950s, and their tow paths are almost gone. With the rising waters, not even their stone walkways will remain to mark their existence. Hersey's story, on the surface, seems to be a simple tale. It might suggest a tale of progress, where the new replaces the old. It might represent loss and gain, for loss and gain, like yin and yang, are always intertwined. The life of Old Pebble was harsh and full of hardships, and Hersey does not make him an especially well-developed or a very likeable character, but with all my heart, I did not want him to perish. I did not want the story to be about the inevitability of change, but I suspect that is what Hersey meant.

Colonies of golden monkeys enjoyed a winter feast in several spots along the shore. Food had been put out for them, and they gathered on the banks like a scattering of gold coins. To see monkeys in a climate where the temperature was so cold seemed odd, but our guide informed us that the summer months were hot and tropical. In a small cove, a pair of mandarin ducks floated among the grasses. Their vivid markings were like a crayon-filled picture from a child's coloring book. With white eye-stripes, orange and cream crest feathers, and white slashes on blue-green iridescent wings, all of their colors were carefully drawn between the lines. In contrast to the vivid mandarin ducks, drab brown grebes fed in the coves, diving down and popping up like cork floats. A fellow passenger observed two tawny fish owls perched on an outcropping of rock. These owls are an endangered species, and with the completion of the dam and subsequent habitat loss, they will probably become even rarer. His wife spotted a common kingfisher. When they showed us the illustrations of the birds in their field guide (Mackinnon and He 2000), I was disappointed to have missed the kingfisher, for it was a lovely green, silky bird.

River dolphins, known as Baiji dolphins, supposedly inhabit the waters of the region, but they are extremely rare. Once revered as water goddesses and symbols of good luck, these dolphins were caught and eaten to near extinction during the famines of the late

1950s and early 1960s. In the failed collectivisation called the Great
Leap Forward, millions of Chinese starved to death. There were
never many dolphins originally—estimates of around 6,000 in the
1950s—but today there may be as few as a dozen. Even though
the State Environmental Protection Administration has spent billions in clean-up projects (Young 2002), the water quality of the
Yangtze River continues to deteriorate, and the dolphin populations are unlikely to recover. In a dark pool at the bottom of a deep
cleft, I noticed the flicker of a fin and hoped it was a dolphin. More
likely, it was one of a dozen species of catfish common on the river
or at best, a rare sturgeon.

In the last Lesser Gorge, the Emerald Gorge, lush bamboo
thickets grew high along the banks. Here the boat turned around
and returned to Wushan. On the way back, we noticed a quaint
village perched on a picturesque hillside. A road ran through the
village, and the houses on both sides were constructed of sturdy red
brick with grey tiled roofs. The homes were surrounded by lovely
trees and shrubs, and birds flew all about the villas. The whole
village had an idyllic air. Among the trees, I noticed a leafless one
with white berries glistening like pearls. Could this be the Chinese
tallow tree? Better known in my part of the world as the popcorn
tree, it has become a troublesome invasive. Here in its native home,
it seemed in harmony with its environment. The fact that many of
the West's beloved ornamental plants originally came from this
region of China, from crape myrtle to ginkgos, from hostas to azaleas, speaks to the great beauty of China's diverse botanicals, but the
popcorn tree, a tree brought to the United States for its beauty as
a landscaping tree, has became a nuisance species in the Gulf Coast.

Returning to the ship, we departed Wushan and continued
our passage upriver. Mr. He explained that the region was rich in
ancient history. He pointed out an archaeological site on the southwest bank where a Daxi village once stood. He said that
archaeologists were working as fast as they could to unearth the
artifacts of this Neolithic culture. With the completion of the dam,

many of the sites would be flooded, and the mysteries of their lives
might never be revealed. Mr. He said that one of the oldest hominid fossils, called Wushan Ape Man, was discovered nearby at a
place called the Damiao Longgu Ruins. The bones carbon-dated at
two million years, making it the oldest hominid fossil in Asia. It
seems rivers have always provided us and our ancient relatives with
good homes.

In the afternoon, we quickly passed through the last of the
Three Gorges, Qutang Gorge, and the remains of the afternoon were
a parade of river cities. The cities of Fengjie, Yunyang, and Wanxian
moved by like tinker-toy towns. Never had I seen so many high-
rises. All those population statistics that had seemed like just large
numbers were transformed into the reality of concrete dwellings.
One-third of China's population lives along the Yangtze. One in twelve
of all the people on earth live along this river. The massive cluster of
buildings was staggering, the sheer construction of it all, overwhelming. All I could think of were ants. Visions of fire ants, their mounds
covering vast fields in loose brown piles of dirt, filled my thoughts as
we drifted by the huge urban centers.

Too soon, it was night again, and the river faded into a black
screen broken only by a few shore lights. With the loss of view came
the evening's entertainment, Mah-jongg, that game of stones and
cards that provided the setting for Amy Tan's mother-daughter stories
in *The Joy Luck Club* (Tan 1989). On the surface, the game appeared
to be simple, but like so many Chinese inventions, it had a complexity that grew as one's understanding of the game deepened. The
game, supposedly invented by a crew aboard a fishing junk in order
that they might forget their sea sickness, requires concentration. I
was tired and my attention wandered, so I turned in early and drifted
off to sleep with images of the emerald waters of the Daning as my
last thoughts.

We awoke the following morning to a quiet ship and realized
we were docked. From our cabin window, the outline of a pagoda
appeared. As the sky lightened, the details of the temple emerged.

Twelve red roofs of the Red Pagoda of Shibaozhai snuggled up against the rock cliffs. Built in the 1700s, the pagoda represents one of the most beautiful examples of architecture along the river. Rather than explore the pagoda, we decided to stay onboard the ship and watch the comings and goings from our warm room. Our room with a view provided an excellent vignette of river life.

The walkway leading from the ship to the shore was lined with make-shift stalls. Everyone that disembarked had to weave their way through the alley of tents. The ship's crew slowly climbed the hill towards town (known as Shibao Block), while the passengers diverged to the entrance of the pagoda. A trio of farmers made their way down from the hills, loaded with vegetables hung in two swinging baskets balanced over their shoulders by a bamboo yoke. After bargaining with the ship's kitchen crew, they exchanged the greens for cash. A group of vendors cooked breakfast by stirring vegetables in a hot, oily wok. Eating rapidly, they washed up and tossed the rinse water into the river. An old legless man crawled down from the high road to the docks, took out his tobacco box, and lit a half-smoked cigarette. He took a few puffs, pinched it out, and returned it to its box.

On the hill among the weeds and garbage, three white goats grazed and roamed about. One of the young ones persistently tried to eat a plastic flag tied to the handlebars of a motor bike. Two fellows grabbed the goat, lifted him off the ground, and strapped him to a pole. They bargained with someone from another ship, and the goat soon disappeared. The goat bellowed in great protest as he was carted away, probably to become their dinner. We watched our fellow passengers appear and disappear in the dense vegetation that engulfed the pagoda. Above the temple, a flock of five cranes flew in V-formation. Too far away to see any clear markings, I wondered if they were Siberian cranes. These endangered birds winter in nearby Dongting Lake and farther downriver in Lake Poyang (near Wuhan). The International Crane Foundation (2003) reports that the winter homes of these rare cranes may be severely impacted with the completion of the Three Gorges Dam and the rising waters.

From the direction of town came two sedan-chairs swaying with two very heavy Asian women. The largest woman, clad in shimmering red and gold, was so vivid everything around her paled in comparison. Her size, attire, and demeanor demanded the focus of everyone's attention. The tiny thin men who carried the women struggled to keep the chairs upright as they wobbled down the slope. To my amazement, they made it down without mishap. The women paid the men, and with their gigantic packages in tow, they waddled down the boardwalk back to their ship.

Since it was winter, life on the docks was not as intense as in the description by Chetham (2002). In summer, when everyone comes to the river to cool off, the riverbanks buzz with activity, but even on a cold, gloomy winter morning there was much to see. When our ship's passengers began returning, they toted bags full of bargains from the trinketsellers. One crisp-looking crew member in his starched white jacket handed the old legless fellow some money. His act of generosity was a comforting gesture as we ended our morning viewing of river town life.

In a few short years, upon completion of the dam, the town would be submerged, although the pagoda would have a dyke built around it to protect it as an important historical site. Supposedly, the people's homes would be moved farther uphill. These facts seemed of little relevance to those working the docks, for life appeared to go on as if there was nothing to worry about. While reports of many large and even violent protests by the local people have been widespread, the docks that morning seemed quiet. One report of a demonstration (McDonald 2004) described a rampage of ten thousand people who had been relocated from the riverside in the Wanzhou township. Unable to find work and living on a meager compensation stipend that would end in the next year, they made their despair known. But today, there seemed to be little indication that the waters would be rising and no sense of urgency that the flood waters would soon cover the town of Shiboa Block.

Back on the river, Fengdu, the Ghost City, drifted by not so much as an apparition but as an actual ghost town. Hundreds of concrete buildings stood abandoned along the lapping shore. The fact that these buildings would soon be dismantled and moved brick by brick across the river to higher ground was staggering, but that was the government's plan, and some work was evident by the piles of debris and partially torn down buildings.

Fengdu is an ancient Ba city dating back 2,300 years. The ghost legends may have begun at a time of great turmoil in China's history, when the Han dynasty was breaking up. The story goes that two early officials moved to the city around 200 B.C. to practice Taoist teachings. Their names sounded like "Kingdom of Hell," so the name stuck. From then on, legends of ghosts and the afterlife seemed to grow and have remained associated with the city. Since over a million tourists come to Fengdu every year to participate in the occult, fortune telling, and supernatural lore, I concluded that there was probably a lot of room for such superstition in the Chinese world view. Tiziano Terzani's odd travelogue, *A Fortune-Teller Told Me: Earthbound Travels in the Far East* (1997), explores this interest in the supernatural in the Asian world.

Near the city of Fuling, large twenty-gallon clay pots covered with weavings were piled along the banks. Mr. He explained that they were used for pickling a mustard tuber. Peter Hessler (2002) in *Rivertown: Two Years on the Yangtze*, describes the hot pickled roots of Fuling as the town's specialty. His account of a two-year stay in Fuling from 1996 to 1998, with all its beauty and brutality, is probably the best written chronicle of modern Chinese river city life. The portrayal of his students' performance of *Romeo and Juliet* and *Don Quixote* was tender; his report of the macho drinking bouts and senseless competitive drinking among the men, disturbing; his depiction of the pettiness of Chinese Party officials, troubling; and his tale of a crowd in Fuling that turned into an angry mob, terrifying. His later book, *Oracle Bones: A Journey Between China's Past and Present* (Hessler 2006), continues the portrait of a modern China going through enormous transformations.

The afternoon passed in leisurely river watching. A plumbeous red start landed on the deck and entertained us with a hopping, wing-spreading dance. The river widened and became a lazy plain of silver washed with islands and mud flats. Now that its waters were less confined by stone walls, the river was a different place, a meandering unhurried place. Small boats drifted on the river and sampans lined the shore. White ibis and plovers fed in the shoals, while gulls, common terns, and fork-tailed swifts flew back and forth across the stern. Herons stood on the flats like ivory sentinels. Sunset turned the sky and surroundings a pale butterscotch. This was the kind of river scene Mark Salzman (1984) depicts in *Iron and Silk*. In the early 1980s, he lived in Changsha, a river town located on a southern tributary of the Yangtze, and his memoir describes his time on the river.

A talent show provided the final evening of ship-time camaraderie. The entertainment began with karaoke, that Japanese invention whose name means "empty orchestra." I am not sure why karaoke has such appeal, but the musical pantomime appears to be as popular in China as it is around the world. After a few songs by our fellow passengers, the staff and crew performed some traditional ethnic dances. China's fifty-six ethnic minorities each has different costumes and dances. When the announcer called out the names of the various ethnic groups, the words did not register. Like so many Chinese phrases, they seemed to float around me like a wispy breeze. My western ear, not attuned to the subtleties of the Chinese language, just could not hear them to remember. As I watched the young men and women whirl and stomp in their colorful costumes, I thought, if they were dancing in the United States they might be dancing the Sun Dance of the Lakotas or a Polish polka or a Cajun two-step. To my unfamiliar eye, the dances seemed similar, but probably like the language, it was a matter of discerning the subtleties.

In the tradition of saving the best for last, the final performance was the most remarkable. A young man as thin as a noodle walked onto the stage and stood alone in the bright spotlight. He lifted a wooden flute and began to play. In the hands of this young man, who swept floors by day and played music by night, the flute became the voice of the river: mysterious, melancholy, and full of the past. The river song was an old one; it was a song that might have been about how cities come and go; how dams were built and dissolved; how floods came, brought death and destruction, and then receded, offering nutrients as restitution; how dynasties rose and fell; how everything changes, even the river. It was a song about how the river is like humanity, full of tragedy, paradox, small moments of beauty, but always, its nature is change.

CHONGQING

The next morning, we arrived on the muddy shores of Chongqing. To bid us farewell, the ship's crew lined the walkway leading to the parking lot. Walking the planks to a waiting car that would take us on to other adventures, I said goodbye to the young housekeeper who moved with the grace of Nureyev. I said farewell to the young man who played music like Orpheus and to the beautiful young woman who had served me dozens of cups of tea. To all the polite, deferential young men and women working on that cruise ship, I nodded and said the only phrase in Chinese I could manage, "xie xie," which means "thank you" and sounds sort of like "she, she". With one last glimpse of the river, I thought about how the beautiful lands, the villages and towns, the fields of good earth, and the wildlife habitats would be changed. All those wondrous river lands, so much a part of China's past, would be submerged by the completion of the dam and the rising waters.

We climbed into the waiting car and sped off into the smoggy streets of Chongqing. This river town, sometimes referred to as the city of a million steps, is built on steep mountainsides, so there were no bikes. Everyone but the few who could afford a car walked, and things were transported by porters known as stick-stick men. For about a yuan—a little more than ten cents—a stick-stick man would carry your bundles. Through the busy market district where the porters trudged uphill carrying the packages for the upcoming New Year's celebrations, we drove in the tour company's black Mercedes. Up the steep hill we climbed, finally stopping only when we reached the top of the mountain and the Three Gorges Art Exhibition Center. The museum housed a mural of the Three Gorges by an artist named Liu Zuozhong. The mural stretched through half a dozen rooms, down hallways and around corners, ending in the ever-present gift shop where a printed brochure of the hundred yards of artwork could be purchased. It never ceased to astonish me how everything the Chinese did seemed mammoth. And there was always a gift shop at every stop.

The thing about Chongqing almost impossible to grasp is that it has a population of over 15.3 million. With the surrounding towns and villages included, the numbers are estimated to be close to 33 million. That is more people than in the state of Florida. When we looked out over the city from the overlook, the fact became, if not comprehensible, at least congruent with the view. The sprawl of high-rises partially wrapped in smog spread out for miles. I had never seen anything like it. The sheer size of the city was shocking. The other river cities did not even come close to this mass of human construction. We were inside the biggest fire ant nest in the world. Far below, the Jialing River curved through miles of buildings like a brown strip of masking tape, while the Yangtze River, lost in the smog and high-rises, completely disappeared.

Also called the rust city because of the acid fogs, acid dust, and the black rain of airborne industrial ash, Chongqing may be destined for yet another environmental impact as the Three Gorges Dam reaches completion. One especially articulate article, "China's Three Gorges Dam—Eco-boon or Cesspool?" (Earthwatch 1997), describes a city that would sit on the shores of an immense waste dump. Since over three thousand industrial and mining enterprises along the river release ten billion tons of wastewater every year, containing more than fifty different pollutants, the reservoir resulting from the dam would become a stinking lake filled with raw sewage and industrial chemicals. Sewage would bubble up, and the city could become the biggest sewer in the world.

At the People's Assembly Hall, we strolled through the auditorium as our guide described the architecture. Far more interesting were the dancers in the square below. Some years ago, a news article reported the young people in one Chinese city were concerned that their parents were staying out late at night dancing and having a good time. They felt their elders were being irresponsible. They wanted them home and in bed. I smiled as I watched the retirees swirling their batons streaming with bright banners. As they spun and whirled their colorful flags with obvious pleasure, I thought, if I lived in this city, I might be one of those late-night elderly rowdies. With such dire predictions of future life among the sewage, maybe they were right to dance while they could.

No trip to China would be complete without seeing pandas, so a brief stop at the city's zoo provided a glimpse of them. The red pandas moved about their cages like hennaed raccoons, while the giant pandas slept folded as giant yin-yang teddy bears. Pandas represent one of China's greatest conservation efforts, and much time, energy, and resources have gone into preserving these endangered animals. China's State Forestry Administration (SFA) and the World Wildlife Fund (WWF), whose logo is actually a giant panda, have worked together since 1979 to create panda reserves. In the bamboo forests of the upper Yangtze, the world-famous

Wolong Panda Preserve conducts research, habitat preservation, and breeding programs on the giant pandas. The preserve has the world's largest population of pandas, and surveys by the WWF report that as of June 2004, about 1,600 giant pandas exist in the wild. Much has been written on the biology and conservation of the pandas, but *Giant Pandas: Biology and Conservation* (Lindburg and Baragona 2004) and *Last Panda* (Schaller 1994) are probably the best.

Our next destination, a Chinese foot massage, began with a pail of warm red water placed at our feet. In went the toes. Within minutes, they were warmer than they had been the entire trip. We were given paper cups of something to drink that looked a lot like the same red toe-soaking water. I figured if it worked on the toes, it might work on the stomach, so I drank it down. Then the boys, who may have been in their twenties but could just as easily have been forty, began to scrape our feet with what looked like straight razors. A little apprehensive at first, I did not feel a thing as the calluses were carved away like thin slices of meat. Then they began to work the muscles, muscles I did not know existed, and for thirty minutes they kneaded our digits like dough. When the massage was done, we laughed and exclaimed that now we had happy feet. And it wasn't an idle claim, as we danced out of their shop.

The drive to the countryside along a narrow road under construction—the normal state of most roads—passed through fields covered with a few inches of water. At a bend in the road, a tiny woman carrying a huge basket of grass on her back smiled and waved. The driver stopped and asked if she lived nearby. She said, yes, and she invited us to her home. Her house was a typical adobe farmer's house of mud and straw with a closet-sized kitchen and two rooms. She placed the grass bundle on the ground, went into one of the rooms, and brought out her prized possession, her water buffalo. As she fed the grey beast the grass she had collected, she told us a little about herself. She had two daughters who lived nearby. She wove baskets in winter and worked the fields the rest of the year. She and her husband grew rice, and she showed us the rice bin where they stored it to keep out the mice. In a dark corner in the tiny bedroom, lit by a single bare light bulb hanging from a single wire, the rice bin looked like a small dumpster. I could not imagine how hard they worked for a bin full of rice.

Her demeanor brought to mind the stories in *The Good Women of China: Hidden Voices* (Xinran 2002). In the 1980s, Xue Hue Xinran, a radio talk show hostess, broadcasted a call-in program called "Words on the Night Breeze." Women called in anonymously and told their stories of hardship and abuse. Some of the narratives were published twenty years later. The old woman could have been one of the thousands of good women of China whose testimonials of painfully restrictive lives were acknowledged.

It was late in the day when we left the farmer's home. Our driver and guide dropped us off at the airport and being young men with things to do, zoomed off into the dusk. As we sat and waited for the flight to Xian and more sights to come, I felt awed by all I had seen on the river. Yet the gloomy predictions of Chongqing's scatological future and the effects of the dam weighed heavily in my thoughts. China was such a vast, unfamiliar time and space, a language, a history, a world so different from my own. But the river was a connection; it had reminded me of my childhood when I read Pearl Buck's *The Good Earth* (1932) and believed the Chinese were a people who valued their ties with their land and rivers. I thought how for thousands of years humans had lived along China's rivers, cultivating the rich river lands and fishing the waters. And in less than twenty years, the Yangtze had been transformed into a river girdled in concrete and steel and polluted with industrial waste. It was a river dammed, a river squeezed with skyscrapers and gargantuan cities, and a river traversed by supertankers. Perhaps the river would survive all the impacts of modern China; perhaps all the things that could go wrong with the Three Gorges Dam would not; perhaps the river would remain great even when humans were finished using it—the way our technological society uses natural resources with such indifference. Like the mystery that rose from the flute played by the young Chinese house-keeper, my thoughts were full of . . . *perhaps*.

Ganges River: One Lone Ibisbill and a Half-Billion People

FROM DELHI TO NAINI TAL

The congestion on the Delhi roads was a tangle of people—on foot, on bicycles, on motor bikes, stuffed into buses spewing white smoke, in auto-rickshaws, in cycle-rickshaws, in little white cars called Ambassadors made by Hindustan Motors—as well as cows, dogs, and pigs, all crammed together and oozing along like a muddy river. Before leaving for India, I read a number of travelogues in an effort to prepare myself for the Indian experience (Naipaul 1991, Mishra 1995, Watkins 2002, Macdonald 2002, Mehta 2004), but nothing could prepare me for the overwhelming mass of humanity and the impact of that mass on the landscape.

It took three hours of stop, go, and crawl to leave the city. The sky was school-hall grey, and the smog from an industrial complex on the outskirts of the city was so dense and noxious that it burned my throat like acid. Once out of the city proper, the road congestion expanded to include brightly decorated Tata trucks, tractors carrying enormous loads of sugar cane, carts drawn by water buffalo or horses, beat-up old minivans, and rural buses crammed so full of bodies that not an inch of spare space remained. Practically anything with wheels that could be ridden was used for transportation. The roadside menagerie included not only cows and dogs but monkeys, camels, goats, sheep, water buffalo, horses, all packed onto a lane not much wider than a driveway.

It took fourteen hours of dodging trucks and carts, darting in and out around slower-moving vehicles, and weaving around obstacles with an overnight at Ramnagar, to travel two hundred miles from Delhi to Naini Tal. Our route through the flatlands known as the Gangetic Plain and then up into the foothills of the Outer, or Uttaranchal, Himalayas was a geographically diverse one. A wide and thirsty stretch of land lying below the giant upheaval of the Himalayas, the Gangetic Plain holds the three great Indian-Pakistani rivers: the Indus, the Ganges, and the Brahmaputra. They fan out like fingers in three directions, clawing their way down from the roof of the world to the Indian Ocean. The Ganges, which begins in the Himalayan ice cave of Amarnath in the northern states of Jammu and Kashmir, flows south and then curves eastward across northern India. We would travel in its upper watershed, around the Kosi River tributary, return to Delhi, and travel south to where the Ganges waters bend eastward.

About twenty miles outside the city of Delhi, the agricultural lands began as flat plots of sugarcane. Bordered by mango trees and tall, gangly eucalyptus, the cane fields were green cubicles. An ancient crop probably first cultivated in the Indus Valley, sugarcane dates back at least 2,500 years, and little seems to have changed in the labor-intensive way of harvesting it. All stages of cane harvesting were visible: some fields had sprigs only a few inches high; some had grassy plants over ten feet tall; some fields were newly cut; and some were burning. The sweet cloying odor of burning cane brought back memories of cane harvest in Louisiana. As a kid, I loved the smell of burning cane fields and the hard sweet chew of a freshly cut stalk. In the cut fields, the stalks were piled high onto carts and carried to sugar mills scattered along the road. When fully loaded, the cane carts resembled bloated dirigibles, and passing them on the road could have been a crisis, but our driver, a calm and skillful fellow, maneuvered around them like a pro.

Through town after town—Hapur, Gajraula, Amroha, and Moradabad—in an endless stream of traffic and people, we rode. The roadside seemed to be a continuous pile of garbage, like driving through a landfill. Trash, dung piles, dung paddies drying in the sun, and dingy stalls selling bags of potato chips, packages of sweets, bottled drinks, and eggs lined the road. Long stretches of hovels, shacks, and makeshift shanties of tarps thrown over poles created a mural of endless brown grime.

Where the road crossed the Ganges River at Brij Ghat (*brij* translates as "bridge" and *ghat* means "stairs leading to a river") we made our first stop. From atop the bridge, a flock of ruddy shelducks formed a grainy amoeboid blob on the surface of the river. Scanning the waters in hopes of seeing a river dolphin, we observed only plastic bags, garbage, and other flotsam bobbing along the coffee-milk colored waters. Eric Newby (1966) in his witty travelogue, *Slowly Down the Ganges*, describes a river strewn with the remains of funeral pyres, but thankfully, we did not see anything that resembled human bodies. Boatloads of tourists and pilgrims motored or poled slowly from bank to bank. A fellow in a brightly-painted blue boat splashed his arms and legs with river water, then dipped an empty bottle into the river to fill it, collecting sacred water to take home, I guessed.

The land around the river was flat and treeless. With so little relief, the smog blurred the line between land and water, and it was difficult to tell where the river ended and the shore began. Downriver, a muted line of shanties hugged the shore. This was where the cremation ceremonies took place. The smoke from the cremation fires drifted and blurred the landscape so that everything had a smoky patina. It could have been an abstract landscape painting by William Turner. How beguiling Ganga Ma seemed from our vantage point, shrouded in its cover of haze. But the scrim did not hide the reality of the river. I recalled some of the things that entered the river (Alley 2002): a billion liters of raw sewage a day, burnt and partially burnt human corpses, livestock corpses, factory wastes of all sorts, industrial and agricultural chemicals, oil refinery effluents, the washings of the ten million people a day who bathed in it. This river, considered a goddess by Hindus, believed to be holy, to have the power to purify everything it touches, and to aid the dead on their path to heaven, is one of the most polluted rivers in the world. Stephen Alter's travel memoir, *Sacred Waters: A Pilgrimage up the Ganges River to the Source of Hindu Culture* (2001), tries to explain the interconnections of the river and the people's beliefs, but the reality of a river considered sacred, yet allowed to become a cesspool and a vector of disease, seemed a great paradox.

For over 1,550 miles the river rolls across northern India, collecting all the waste of a human population estimated to be 350 million, or one in every fourteen people on earth. Where it empties into the Bay of Bengal, it unloads its sludge in a swamp-forest known as the Sunderbans. This delta of mangrove forests, crocodiles, and man-eating tigers, depicted in the novel *The Hungry Tide* (Ghosh 2004), is a place of great beauty and mystery. What by all modern definitions is a sewage terminus becomes in the hands of the novelist a place of vivid imagery and mysterious happenings. Like *The River Sutra* (Mehta 1994), another lovely novel about another heavily impacted river, Indian literary writings on rivers seem full of contradictions. For all the beautiful words, the sacred beliefs, and the mythology of the river, the truth is that India's rivers are heavily polluted, and the Ganges is one of the most heavily impacted rivers in the world.

For twenty years, a group of concerned citizens has been trying to clean up the river (Stille 2003). Under the leadership of V. B.
Mishra, the Sankat Mochan Foundation has organized workshops,
launched educational programs, and tried to build support from
government and industry. A recent Clean Ganges Campaign had
some success, but the sheer magnitude of the problem makes it a
daunting task. Efforts to pick up litter along the river and to build
effective wastewater treatment facilities have resulted in some
progress, but it is an uphill struggle.

Another organization called Navdanya, founded by the in-
domitable activist Dr. Vandana Shiva, has also been involved in
efforts to protect the Ganges. Dr. Shiva, as described in her *Water
Wars: Privatization, Pollution and Profit* (Shiva 2002), has been fighting
multinational corporations that want to privatize the Ganga waters. Her efforts in water rights as well as in biodiversity conservation (keeping the traditional and sustainable mustard as a crop rather
than adopting the corporate-controlled soy bean), farmer's rights,
and antipoverty programs are part of a larger effort at improving
the lives of the poor. Navdanya and Dr. Shiva's Research Foundation for Science and Technology and Ecology are both public
scientific attempts to protect the Ganges. Protecting the river is the
theme of Arundhati Roy's elegantly written essay "The Great Common Good" (Roy 1999), which addresses the issue of who owns the
river and its waters. She and other activists have been protesting
the Sardar Saravar dams proposed on the Narmada River for many
years in an effort to protect the river and the people who live on it.

As we stood overlooking the Ganges, a woman of undeter-
mined age began a conversation with us. We could not understand
a word she said, but our guide, O.P., translated her monologue to
say that no dolphins had been seen in a long time. Her hair, a mat
of grey tangles, framed a wrinkled, milk-chocolate face, and her dark
sunken eyes were the color of soot. Her earlobes were holes that once
held earrings, but now only a tiny silver nose stud remained, perhaps
the last remnants of waning beauty and dwindling wealth. Her voice

was high-pitched, like the sound of small bells. She had a few crooked teeth on top and fewer on the bottom, but she smiled broadly when we took her picture. When I gave her a hundred rupees (a couple of dollars), she rushed off in such a hurry that I wondered if she had urgent business or if she was trying to slip away quickly without drawing undue attention to her newly acquired wealth. In her travelogue *East Towards Dawn*, Nan Watkins (2002) was warned not to give "so much as a single rupee to one Indian beggar" or she would have an army of beggars following her. But I simply could not resist giving her what to me was little more than pocket change.

A shriveled old man wrapped in white cloth hobbled past us carrying a walking stick. He looked like Gandhi. As he wobbled by, I thought how much our image of India had been shaped by Gandhi and how much he and India were entwined. The film *Gandhi* made such an impression on me, and the story of the 1930 salt march is one I hold as an icon of great courage. Marching 240 miles from his home on the Sabarmati River to the sea in protest of the British salt tax, Gandhi's act of civil disobedience was one of many aimed at creating an independent and socially just India. His life defined him as one of not only India's but the world's great spiritual and political leaders.

I wondered why so many spiritual leaders seemed to originate from this country. The media are full of stories about them. In *Karma Cola: Marketing the Mystic East*, Gita Mehta (1994) presents a humorous look at some of them, and in the popular spiritual travelogue *Eat, Pray, Love*, the author (Gilbert 2006) describes her stay in an Indian ashram where her unnamed guru was not even present. One of the spiritual celebrities known as Amma, or the hugging saint, has been described in *Amma: Healing the Heart of the World* (Cornell 2001). By last count, this mystic had hugged ten million people. Others with strange and complex names like Bhagwan Sri Sathya Sai Baba, Dadi Prakashmani, Gurumayi, Sri Ramana Maharshi, Swami Adiswarananda, and the list goes on, have frequently appeared in the news. In the late 1960s, the spiritual leader known as the Maharishi Mahesh Yogi enlightened the Beatles, and in the 1980s, the infamous Bhagwan Shree Rajneesh made headlines when he moved from India to Oregon and got in trouble for tax evasion, as well as irritating the locals with his proclivity to ride around in a new Rolls Royce every day (he had ninety-three of them). I wondered if these charismatic people had the same attraction as our own David Koresh of Waco, Jim Jones of Guyana, Jimmy Swaggart, Jim Bakker, Pat Robertson, Pat Buchanan, and the list goes on. Of course, the question arises: Are they individuals whose actions come from a self-serving need for power, or are they legitimate spiritual/religious leaders genuinely interested in making the lives of people better? I cannot answer that question for those in India, but I think in the United States it is quite clear what motivates many of our religious celebrities.

Our guide O. P. on several occasions during our travels tried to explain his religious beliefs. He explained that he was Hindu, a Brahmin, and that his father had been an influential Marxist leader in his home town. He said that his family believed him to be spiritually inclined, but I had a difficult time understanding his philosophy. He claimed that the world's oldest religions, Hinduism and Buddhism, began in India. Then he went on to explain some of the beliefs of Hinduism, but when his explanations began to sound like proselytizing, I lost interest. As someone educated as a scientist, I find religious discussions rather uninteresting most of the time. As someone from a world of comfort, convenience, and great personal freedom, it seemed a paradox that such compassionate forms of enlightenment could originate in a place of grime, poverty, and minimalism. Yet it is possible that religion arose more as a means of escaping the brutality of life and maintaining dignity in the face of poverty and pain than of trying to make sense of the world.

Of all the sights on our days of highway travels, the image of a well-groomed elderly gentleman sitting on a rusty old bike, pedaling his way to work among the exhaust fumes, the mangy cows, the garbage, and the desperately poor people seemed to best fit my idea of dignity. I would have liked to have seen an elderly woman doing this, but there were few women driving anything. I noticed no women drivers on our travels. The women in cars, buses, motorcycles, or other vehicles were always passengers. There were no women shop keepers, hotel workers, cooks, cleaners, or waiters. With the exception of two women clerks behind the desk at a hotel in Delhi and a few in the women's toilets handing out squares of paper, I saw no women working in jobs in the public realm. Women seemed to be elsewhere and out of sight. The women I did see along the roadside were either squatting in the fields working the soil, shopping at food markets, or making dung paddies. The sight of one young woman squatting along the roadside with her hands deep in a pile of dung was rather depressing. But when I noted the expression of such utter hopelessness on her face, it was deeply distressing.

When O.P. said in a rather superior tone that "making dung paddies was women's work; men did not do it," it was hard to control my annoyance. The days of dust and grime, the constant honking of horns, and the mixtures of cooking smells and decaying matter with all the impact of travel was affecting my ability to maintain a nonjudgmental attitude. Sensory overload seemed to be wearing my emotional control thin. Throughout the journey, I would recognize a discomfort I had not felt before in all my travels. Not only was I a woman traveling in a foreign land, but I was a woman without the casual interactions and pleasant contacts with other women. I have always enjoyed such exchanges, but during this journey there would be few encounters with women. Sometimes a journey can be defined by what you don't experience.

NAINI TAL

The road from Ramnagar to Naini Tal was greener with more pleasant farmlands. Crossing a river that closed the road when the monsoons came, yellow fields of mustard and orchards of mangos stretched out against the blue mounds of the Kumaon Hills. Farm yards full of marigolds and patches of wild lantana added splashes of ocher to the scenery, and birds were everywhere. On the wires, black drongos balanced with long trellised forked tails, bank swallows swooped among the trees, and ashy-throated bulbuls nestled in the branches. On a single tree, seven steppe eagles perched like enormous markers, one for each day of the week. Half a dozen weaver nests hung like straw teardrops from a eucalyptus tree, and green bee-eaters, in sharp contrast to the dull-colored weaver birds, balanced on the wires like jeweled stick pins.

Almost before we could enjoy the loveliness of the country-side, the road began to climb, and for four hours, our driver made hair-pin curves as we rose into the Himalayans. With the higher elevation, the dusty plains gave way to greener forests. Tropical dry deciduous species and sal trees (*Shorea robusta*) made up the lower forest, then gave way to chir or pine forests (*Pinus wallichiana*). Evergreens like cedars (*Cedrus deodara*), fir (*Abies pindrow*), and spruce (*Picea smithiana*) appeared. Oaks (*Quercus incana, Q. leucotrichophora,* and *Q. dilatata*) and other broadleaf species, with dense thickets of rhododendrons (*Rhododendron arboreum*), also greened the slopes. By five thousand feet the moist temperate forest of the Himalayan foothills was an array of evergreens and mixed deciduous trees, not only oak but elm, alder, willow, poplar, horse chestnut, hickory, and birch. These were the forests of the first tree huggers, known as the Chipko movement, where village women and children formed circles around the trees to protect them from government contract cutters (Weber 1988).

The hills were extraordinarily steep, and the sky changed from the dull haze of the lowlands to a deep azure. Above a stand of pines, a lammergeier soared in the clear mountain air. It was the first vulture we had seen, and as we watched it glide against the mountainside, O.P. told us how the vultures of India were disappearing. For the past few years, fewer and fewer vultures had been reported in the region, and no one really knew why. I wondered if the myriad of industrial toxins had finally made it up the food chain to the vultures. Or maybe the birds had just had enough of India (I was letting a few sarcastic thoughts slip into my mind).

At a roadside stream, we paused to stretch and enjoy the cool mountain air. A pair of white-spotted forktails bobbled in the rushing water, and we watched their antics for a while. Moving on to Naini Tal, we entered a resort town. Once a small hill station, the town had bloomed into a vacation place for the affluent who could afford to escape the summer heat and humidity of Delhi. Naini Tal (which means "eye lake") had over two hundred hotels and hundreds of guest houses and cabins. Up steep and narrow lanes, our driver steered through the stalls, around the government complex and courthouse, and finally to our hotel, where tea and biscuits took the chill off the cold evening air.

Twelve miles from Naini Tal, a forest preserve at Sat Tal (seven lakes) offered us the first chance to hike. The trail began near a Methodist church camp, where a flock of slaty-headed parakeets congregated in the branches of a plum tree. With their long tails hanging down, they looked like green icicles. The rocky path sloped down to a small algae-green lake and looped through a forest of moru oaks (*Quercus dilatata*). In the sunlight, the leathery leaves glistened like hand mirrors, and candelabra cactuses speckled the slopes. What a strange mixture of flora and fauna in this mountain habitat, the combination of oaks, parakeets, and cactuses. The next lake over was the color of cocoa, and from the opposite shore we heard the strident call of "kay oh," a great barbet making its presence known.

Nearby the lush green Laxman Campgrounds provided an elevated creek trail in our search for birds. A crimson sunbird flittered in the lantana, sparkled like a trinket, and disappeared so quickly we wondered if we had only imagined him. Two women carrying piles of fodder moved past us on their way to feed their cattle. The grass was piled so high and wide it covered their entire bodies. It looked as if giant haystacks were moving through the forest on tiny brown feet.

For two days, we explored the farm valleys of Pangot and Mongoli. The mornings were cold but we warmed as we walked. On the shady road to Pangot Valley, we encountered a few young men on horseback, villagers on foot, and kids in uniforms walking to school. From the road that ran along the hillside, we climbed down and traversed the valley through the fields where farmers weeded patches of sweet peas, spinach, and white carrots. The haystacks were perfect mounds of yellow grass, and the rising sun transformed the straw into gold. Like gilded hutches, they glimmered in the green fields. The farmers' homes were adobe whitewash with vividly painted red or blue doors. It was hard not to fall in love with this pastoral landscape. Resting in the shade of a cedar tree, we watched a pair of rufous treepies chase one another. From tree to tree, the large birds threw themselves into the branches, dragging their long tails around like towels.

En route to Mongoli Valley, we rounded a steep curve and the picture postcard snow view appeared in the distance. The peaks of the Garhwal Himalayas rose like a white picket fence. The chain of peaks, three of which (Nanda Devi, Kamet, and Trisul) towered over 25,000 feet, seemed to be holding up the clouds. In the Hindu pantheon of gods, Nanda Devi is the goddess of bliss and the reincarnation of Parvati, the wife of Lord Shiva. As the third highest mountain in the Himalayans, Nanda Devi is well named. At over 26,000 feet, its peaks (an east and west one) are noted for mountaineering expeditions and popular sites for Himalayan trekking adventures.

In the valley, we wandered among the hilly fields and watched farm kids chase away a troop of foraging rhesus monkeys. The monkeys were never killed or harmed, just scared away with loud shouts, O.P. explained. This was the Hindu way, he said. How different this view of life was from the Muslim practice described in *Maximum City: Bombay Lost and Found* (Mehta 2004). The sacrifice of bulls and goats during the Muslim festival known as Bakri Id depicts animals twitching and gasping in death throes and blood running like rainwater in the streets. The killings are as graphic as it gets. Then I recalled the reports of violence erupting from fundamentalists Hindus when the feminist filmmaker Mehta Deepa tried to film, *Water* (Pressley 2006) and the chronic Hindu-Muslim communal violence reported in the news. Hindus, Muslims, who believes what? It all seemed far too confusing for me to understand.

CORBETT NATIONAL PARK

The next leg of our journey to Corbett National Park took us into the foothills of the Outer Himalayas known as the Shiwalik Mountains. India's first national park, Corbett is named in honor of Jim Corbett, hunter and author of *Man-Eaters of Kumaon* (Corbett 1946). Corbett's exploits represent the India of English colonial imperialism and the bygone era of maharajas, man-eating cats, and great white hunters. His accounts of tracking and hunting tigers and leopards of the Kumaon hills with his trusted dog, Robin, deal mostly with the details of finding and killing the predators. While he was certainly a hero to the local people of the region who suffered hundreds of deaths from the old or wounded man-eating leopards, his most popular novel is a rather outdated hunter's tale but much of colonial literature is outdated, except in a historical context and hardly in a conservation mode. However, selections from his six novels made by his editor (Hawkins 1978) present a gentler picture of Corbett and reveal his deep love and respect for the poor of India. He writes of his need to hunt in order to eat and his tender care of an injured antelope, among other topics. Like many hunters-turned-naturalists, his efforts later in life resulted in a preserve that today protects many endangered plants and animals.

The road into the park was lined with teak trees (*Tectona grandis*). Like the pine plantations of my home, the monocultured exotics with their tall straight trunks grew at regularly spaced intervals, giving them the monotonous look of a platoon. Leaves the size of placemats sprouted in clusters all along the trunks, and the understory was almost bare. Because the enormous leaves captured every quantum of sunlight, little grew in the trees' dark understory.

We changed vehicles, from the little white Ambassador to an open jeep, and entered the tiger preserve. Bouncing along the red dirt roads, we spent the day searching for tigers. We jostled up and down the hilly ridges as the park naturalist pointed out some of the sights. He explained that the park was divided into a northern portion, known as a moist bhabar dun-sal forest ecosystem, and a southern portion, known as a terai-bhabar ecosystem. A dun is an elongated valley, and a terai is a low swampland; bhabar is the most abundant grass (*Eulaliopsis binata*) and sal is the most abundant tree (*Shorea robusta*). Botanical surveys of the park report high species diversity, with 110 species of trees, 51 shrubs, 27 climbing species, and 33 species of bamboo, with sal forests making up about 75 percent of the woodlands (Singh et al. 1995).

In some spots in the lower hilly ridges and flat valleys, pure stands of sal trees created forests so dense that not even a little grass grew beneath the trees. The tall, straight trunks rose like rusty columns, and beneath them in the dark shade, termite mounds formed red sand castles. In other places, the forest was a tangle of dense shrubs with vines and trees such as the haldu (*Adina cordifolia*), rohini (*Mallotus philippinensis*), karipak (*Murraya koenigii*), and sain (*Terminalia tomentosa*). Occasionally, a smooth trunk Bo tree (*Ficus religiosa*) wrapped around a host tree and created an interruption in the forest. With its long rootlets hanging like ropes and its canopy elbowing its way up for sunlight, the tree altered the texture of the tangled jungle. Like the throne of a powerful queen, it sprawled out and claimed its place in the dense vegetation. A species of strangler fig, the Bo tree is considered sacred by Hindus and Buddhists, for it is believed to be the tree under which Buddha obtained enlightenment. It was easy to understand why such legends surrounded this colossal, magnificent tree.

We forded rivers lined with trees like the shisham (*Dalbergia sissoo*) and khair (*Acacia catechu*). Dense stands of bamboo (many species but most commonly *Dendrocalamus strictus*) and lantana (*Lantana camara*) added to the lush vegetation. A neem tree (*Azadirachta indica*), with its long evergreen eucalyptus-like leaves, grew high up on a sandy riverbank. This indigenous evergreen in the mahogany family has been a traditional source of medicine for villagers for thousands of years. When W. R. Grace and Co., a multinational corporation, tried to obtain patent rights and control the use of the tree, farmers and villagers demonstrated against this act of biopiracy. The tree has become a symbol of opposition to the commoditization of natural resources by multinationals in India (Shiva and Holla-Bhar 1997).

We roller coasted up and down the higher ridges, where trees like bakli (*Anogeissus latifolia*), chir (*Pinus roxburghii*), anauri (*Legestroemia paruiflora*), gurail (*Bauhinia racemosa*), and shrubs (*Clerodendrum viscosum, Ardisia solanacea, Glycosmis pentaphylla, Colebrookea oppositifolia*) grew. We saw no tigers that day or on any of the other days, but other wildlife was plentiful. A monitor lizard dangled from its hollow-tree home like a wall hanging. A parade of small mammals including barking deer, spotted deer, hog deer, sambar, chital, and nilgai antelope stared at us wide-eyed and dissolved into the background as we bounced by. Crocodiles basked in the sun along the riverbanks. Two black-neck storks the size of small ponies waded along the shallows of a rocky tributary, and a stork-billed woodpecker posed on a shady branch like a painted ceramic figurine.

Dhikala, a camp overlooking the Ramaganga River, was busy with visitors when we arrived in late afternoon. A popular site for tiger safaris, the compound was filled with folks in khakis and binoculars. From the campsite on the high ridge, the setting sun turned the grasslands into a copper plain, and as dark descended, a herd of elephants grazing in the grasslands at the river's edge faded into the twilight. The next morning, we were up before the sun and off in the bouncy jeep again to look for tigers in the grasslands.

The grasslands, called chaurs or chowds, were simply spectacular in the dawn light! The sun sparkled through the dew on the grasses and transformed the veldt into a china cabinet full of crystal. Seed heads dripped with moisture and glistened like cut glass. The gracefully curved blades of the tall grasses shimmered like water sprinklers. The savanna grasses (the most common species of *Themeda arundinacea*, *Thysanolaena maxima*, and *Vetiveria zizanioides*) were a birder's paradise. Black francolins, small buttonquails, oriental skylarks, common ioras, zitting cisticolas, and prinias popped in and out of the grasses like popcorn.

For two days, we jostled along the rutted roads but never saw a tiger. We spotted prints in the soft sand on the roads, scratch marks on a scenting tree, or "tangy hieroglyphs" as Padel (2005) describes them, and fresh scat. We even heard a groan-cry identified as a tiger with cubs, but never actually saw a big cat. O. P. said this was the first time in all his ten years of coming to Corbett that he had not seen a tiger. In truth, it did not bother me that much, for I was more interested in the birds and the plants. Supposedly there are about thirty tigers in the preserve, with perhaps less than two thousand in the world. To know that the big cats were there and protected in a world where their habitat is vanishing at an unprecedented rate and where poaching continues to dwindle the population (Quammen 2003, Dinerstein 2005, Padel 2005) was enough for me. I was also glad to know that the guides did not "beat the bushes to roust out the one old toothless tiger in the reserve for the tourists to take pictures," as a delightful Indian women I met on the plane had humorously described some of India's tiger preserves.

From Dhikala, we traveled to the Kumeria Forest Preserve in the Bijrani Range, stopping occasionally along the Kosi River, a tributary of the Ganges, for some easy birding. On a steep bank, a wallcreeper climbed the brown slopes and flicked its red underwings like blinking Christmas lights. All around the river, pied, crested, and common kingfishers hovered in mid-air. Suspended high above the river for minutes at a time, they searched for just

the right shadow on the surface of the river, then plunged down
like spears to snag a fish. In a grove of rhododendrons, a pair of
Himalayan flamebacks played chase among the branches. They
could have been colored by a child selecting the brightest crayons
in the pack, razzle-dazzle rose and pumpkin autumn gold.

Fork Tail Creek was a fairly shallow stream when we drove
through it on the way to the lodge. At the edge of the preserve on
a hill overlooking the Kosi River, the inn had a high view of the
valley. From the garden balcony, the boulder-filled river flowed
through rich farmlands green with rice fields and banana trees.
O.P. pointed to a rock creeping along the shore and exclaimed,
"Ibisbill!" The bird moved slowly among the grey and white boulders, probing for invertebrates and small fish. It blended in so well
with the rocky shoreline that it really looked like a meandering
rock. About the size and shape of an avocet, it had a dark, down-
wardly curved bill like an ibis, a black face, and a high belly band.
O.P. said this male was the only one that had been seen for several
years, although in 2000 there were reports of seventeen birds along
a nearby ten-mile stretch of the Kosi River. No one knows precisely
how many of these rare birds exist in the wild. Their habitat is so
narrowly restricted to the foothills of the Himalayas that they are
much sought after by birders. In fact, birders come from all over
the world in hopes of seeing one. But I had an unsettling feeling as
I watched the stilt-like bird rummaging through the rocks. I kept
thinking how alone it must be, this one bird without the company
of others; perhaps it was headed down that lonely lane of extinction, becoming the last of its kind.

On the second night in the forest preserve, a storm rolled
down from the mountains and rain pounded our tin roof so hard
it sounded like the roaring of a train. Thunder exploded overhead,
and rivulets of water poured in through leaks in the roof. Since the
electricity went out early in the storm, we watched the rain leak in
and onto our beds and suitcases by candle light. As the deluge
continued, I wondered if the road would wash out at Fork Tail

Creek. It looked badly damaged when we had first crossed it. The creek flows into the Kosi River, a river with a history of flooding. Called the river of sorrow because of the great suffering its floods have caused, I hoped the Kosi would not swell and sweep away the ibisbill. By morning the storm had passed and the sun came out. Stepping out of the cabin, we heard loud maniacal laughter coming from the underbrush just outside the door. Maybe someone had been driven to the brink of insanity by the storm, but when we looked, a flock of white-crested laughing thrushes was scratching in the damp ground beneath the trees. Adorned in white head-feathers, they were industriously plowing the ground for insects, cackling away like mad witches. David Rothenberg, in *Why Birds Sing: A Journey into the Mystery of Birdsong* (Rothenberg 2005), describes their song as a Charlie Parker/saxophone laugh.

Fork Tail Creek was passable but only barely, as our driver steered slowly and carefully over the submerged road. Leaving the foothills of the Himalayas, the congested road back to Delhi proved as harrowing on the return trip as the going. Overnight in Delhi, we traveled south through the watershed of the Yamuna River, the largest tributary of the Ganges.

BHARATPUR

One of the world's most notable bird sanctuaries, Keoladeo Ghana National Park (commonly referred to as simply Bharatpur) was once the hunting grounds of the Maharaja of Bharatpur. In the 1850s, diverting the river and flooding the grasslands created a 7,165 acre wetlands that attracted birds for hunters. A stone-carved monument remains in the middle of the park as a record of the hunting trophies. The monument lists the date, who visited "on the occasion of the visit of," and the "bag" number. Visits included such dignitaries as the Prince of Wales and the Duke of Windsor, and one entry had 3,200 birds killed on one day's hunt. In 1956, the park was designated a bird sanctuary, and in 1981 a national park, although the Maharajah retained shooting rights until 1972. It was hard not to make some caustic remark as we viewed the sporting entries chiseled in stone.

For the last five years, the monsoons had not been good, O.P. said, and the shallow lakes and woodlands were drying up. As a result, fewer and fewer birds were coming. No longer were Siberian cranes migrating to Bharatpur. In 1976, seventy-six cranes were seen in the park, but by 1992, only six cranes had come. In 1993, Peter Matthiessen (2001) reported seeing five cranes. Then the number dwindled to a single pair. By 2001, they too had disappeared. When I learned this, I was saddened, for India's loss of Siberia cranes was probably something that would not be fixed.

O.P. arranged for bicycle rickshaws and cyclists to pedal us through the park. The pathways along the dykes were good vantage points to see both the wetlands—the lakes, canals, and marshes—as well as the semi-arid plains. The juxtaposition of wetlands and dry plains exemplified the contrast in landscapes of the region. The marshes held aquatics, some old familiar ones like water lilies (*Nymphaea nouchali*, *N. stellata*, and *N. cristata*), lotus (*Nelumbium* species), duckweed (*Lemna* species), water fern (*Azolla* species), aquatic grasses (*Vallisneria* species, *Hydrilla* species, *Chara* species, *Ipomoea* species), sedges (*Cyperus* species), and lesser reedmace (*Typha angustata*). The grasslands resembled the savannas of Africa and were peppered with babul trees (*Acacia nilotica*), ber or Indian jujube (*Zizyphus mauritiana*), khejri (*Prosopis cineraria*), evergreen shrubby trees (*Salvadora oleoides*, *S. persica*), and the wild caper (*Capparis aphylla*).

We roamed the park for three days, and by the end of the last day our bird list was over a hundred new birds. They included the greylag geese, known to fly over the Himalayas reaching altitudes of 23,000 feet or more (I wondered how they prevented oxygen deprivation and learned that their hemoglobin has a higher oxygen affinity); painted storks, fishing with the seriousness of scholars bent over manuscripts of mud; Sarus cranes that danced like ballerinas; and Indian rollers that flashed turquoise and wild-blue-yonder blue when they took flight.

The final evening in the sanctuary, with the setting sun casting a metallic patina over the landscape, I glanced back along the trail for a final look. A pair of jackals slunk across the path and faded into the brush like bronzed dogs. Village boys rounded up the herds of cattle grazing on the grasslands and began driving them home. On the way back to the guest house, O.P. talked about working with an American ecologist who had come to study the effects of cattle on the park lands. For all the good such studies do, I thought. All over the world, cattle damaged grasslands, and the solution to an economic and ecologically unsustainable problem continues unresolved. Humans seem determined to have as many cattle as possible, whether it's in Africa, Argentina, Australia, or America, and to graze them in ways that damage grasslands. The damage from overgrazing and trampling seems a downward spiral in grassland ecology, and always such devastation seemed to be linked to humans' need for more.

The road to the Bundh Baretha Wildlife Sanctuary was full of potholes, and the ride was bumpy but full of wondrous sights. Through miles of yellow mustard and wheat fields muted by heavy morning fog, through a few dismal villages, we bounced in the little white Ambassador. When the morning mist lifted, the green fields turned to dusty red hills as we approached the dam and reservoir. From atop the dam, the reflected images of the Aravalli Hills and a Buddhist shrine floated in the lake. The hills were red, the shrine was red, the lake was red; the whole world felt red.

Part of an ancient mountain range, the Aravalli Hills are the oldest mountains on earth. What a contrast to the Himalayas: we had traveled from the youngest mountains, with their steep lush terrain, to the oldest, with their dry low-rolling hills. The landscape of the region seemed to have an antonym for everything: the oldest mountains, the youngest mountains; the driest plains, the wettest forests; the most congested cities . . . well, maybe not everything.

Passing a stone quarry, I recognized the red sandstone that made up the great buildings of the region. O.P. claimed the government had stopped mining in that particular quarry because of the impact on the environment, and I thought that seemed like a very progressive thing to do—for a government that does not have an especially good track record of environmental responsibility. We crossed the weir bridge, pulled off onto an ancient stone road built in the days of the raj, and jostled up a steep hill to the abandoned palace. The lovely old building was made of red sandstone and belonged to the family of the Maharajah of Bharatpur. It was going to be renovated and turned into a luxury hotel for tourists. As I looked out over the spectacular panorama of lake, islands, and hills, I believed I understood the "progressive environmental legislation." Tourists would pay big bucks for a room with such a view, but they would hardly want the lake polluted with quarry wastes. If that stone quarry had been elsewhere instead of near a wealthy landowner's property, there would have been no stopping the mining "to protect the environment." But regardless of the ulterior motive, I figured whatever works to make any environment healthier, even if it is motivated by profits for the already rich and powerful, does some good. In this case, it was probably a sustainable economic venture.

Returning to Bharatpur, we pulled into a farmyard and walked to a grove of eucalyptus trees. I thought we had stopped to do a final evening bird watch, but what hung from the spindly branches were dozens of long brown chrysalises. It was a colony of flying foxes. These fruit-eating, tree-pollinating bats, whose heads look like those of little foxes, use sight and smell rather than echolocation to get around. During the day, they hang out in roosting camps like these. Restlessly, the bats pulled their wings around their bodies like shawls. Leather-wrapped, they closed like tent flaps. They would soon be off on their nightly forays for mangos, guavas, and bananas. I wanted to watch them take to the night sky, but we needed to be off the road before dark, and so we left the bats to their awakening.

Back at the guest house, we strolled around the grounds, thinking a short walk would be good exercise before dinner. Pausing briefly under a large banyan tree (*Ficus bengalensis*), I noticed a bee on my shoulder. One landed on my traveling companion's shirt, then another and another, and suddenly we were speckled in bees. We ran like sprinters, leaving the bees behind. Luckily we weren't stung (we decided they were Buddhist bees), but it was a reminder that this strangler fig, with its complex interdependent associates, feeds and houses a myriad of creatures. The tree was always full of hornbills or crows, ants or beetles, bees or wasps. Something was always crawling about in the branches of this accommodating fig tree.

Dinner with O. P.'s family was traditional Hindu cuisine cooked over an open fire. His wife, Phoolo, and daughter made chapatis. To a bowl of flour they added a little liquid butter and water. The dough was folded and kneaded for awhile. Then with the skill of a potter, Phoolo pinched off a golf ball-sized piece of dough, flattened it, and twirled it around like a pizza crust. She placed the thin disk on a skillet and it puffed up. When it was partially cooked and stiff, she removed it and leaned it against the oven wall to brown. She kept the toasted diskettes coming, and over and over we gulped down hot chapatis until we were stuffed. The vegetarian dishes of greens and various beans were also good, but the chapatis were unbelievably delicious.

We learned that O.P.'s daughter was to be married the following month. It was an arranged marriage, and she had never seen her husband. O.P. explained that weddings were expensive for girls. Dowries had to be paid. Sometimes the family of the bride had to buy the family of the groom a car or some other expensive item. I remembered the chilling story entitled "The Firebird's Nest" (Rushdie 2004), which describes men who married for dowries, burned their young wives to death, and remarried to acquire another dowry. The words of Kavita, a character in Gregory Robert's epic novel *Shantaram* (Roberts 2003, 561), came to mind. She exclaims, "but I've got to tell you, Lin, we get too many stories like this every day. Wife burning, dowry murders, child prostitution, slavery, female infanticide—it's a war against women in India, Lin. It's a fight to the death, and mostly it's the woman dying." I tried to shake off those extreme images. I knew O.P. and Phoolo were good parents; they would never allow their daughter to marry into a bad situation.

Yet no matter how hard I tried to be nonjudgmental about these affairs of domestic and social arrangements, I had difficulties with arranged marriages. It was hard to imagine this lovely young woman leaving her childhood home, marrying a man she had never met, and spending the rest of her life in the home of her husband's family. It seemed so restrictive, so oppressive that I must have shuddered when O.P. related the events. He quickly quoted the statistics on arranged marriages versus love marriages, saying that they lasted as long, but it did not make me feel any better. That night, as I lay on the hard mattress trying to coax sleep to come, I could not stop thinking of the words of the old folk song sung by Joan Baez, "Oh, hard is the fortune of all womankind. She's always controlled, she's always confined, controlled by her parents until she's a wife, a slave to her husband the rest of her life."

The final destination of our journey was the Taj Mahal on the banks of the Yamuna River in nearby Agra. It was late afternoon when we arrived, and the pearl white symmetry of the spectacular tomb blushed a pale rose. A monument of love and grief, its splendid dome and minarets reflected the perfection of evening. With such mathematical precision, its structure seemed contradictory to those powerful emotions from which it supposedly sprang. How could something so balanced originate from the most intensely irrational emotions of love and grief? The guide recited the touristy spiel: that it was built by the fifth Mughal emperor, Shah Jahan, to honor his favorite wife, Mumtaz Mahal (I wondered if it was an arranged marriage, and realized, of course, it was); that it took twenty years and 20,000 workmen to complete; that the marble came from Makrana, the turquoise from Tibet, the jade from

Turkestan, and the rare shells and gemstones from all over the empire; that the pietra dura inlays, the hardstone carvings, the calligraphy, and the incised paintings were works of great craftsmanship. On and on, the guide recited the details of its construction.

As I stood enthralled by this stunning mausoleum, I was once again reminded of the great paradoxes of India, the great contradictions in both the country and the region we had traveled. India has one of the finest engineering schools in the world, yet two-thirds of its population is illiterate; it is a land where the forests have been cut and the natural environment decimated by a population of 1.3 billion, yet so many things in nature are revered; it is a place where a civilization as old as any on earth has managed to continue, where great religions and enlightened philosophies have originated, yet most people have no safe drinking water or means of treating their waste; it is a place where the most powerful person in political office was once a woman, and the Hindu religion is full of strong goddesses, yet I saw few women working in ordinary jobs. Near the filthiest, grimmest hovels imaginable stood the splendor of the Taj Mahal. All of these contradictions in the place where the Ganga Ma rushes from the world's highest mountains down to the plains, where the holiest of rivers is also one of the most polluted.

Of all the rivers I traveled, the Ganges most represented that most distinctly human quality of paradox. It embodies the dichotomy of life and death, the duality of being and not being. To be human is to be aware of paradoxes; to be alive is to accept that life is a river of moments that have a beginning and an end. Perhaps the understanding of paradox is what connects us to place, to time, and to each other. And perhaps the Ganga Ma moves like the continuum of humanity, with all its contradictions, with all its illogical and rational ways, with all its selfishness and generosity, its intolerance and compassion, all its hopes and despairs, with all its paradoxes.

Nile River:
Tombs, Temples, and Tourists

Gone are the days when a writer could travel the world and out of those journeys bring forth great truths. Herodotus, Marco Polo, and Darwin were acclaimed as great discoverers in their times because their travels and findings were rare. Modernity offers less opportunity for discovering truths via the journey, because nearly everyone travels and publication is based on marketability. These are the realities. Tourism has become the world's largest industry, with more than 700 million international travelers in 2002 (Osborne 2006). In a time when 100,000 to one million book titles are published each year, the reflective journey is not especially valued, unless it is written by a celebrity or promoted as high adventure, sentimentality, crisis, victimization, or some other sensational quality. *Travel, Discovery, Animal Planet, National Geographic*, and other television channels devote airtime to "exotic places" with sensationalism and more advertisements than information. All these facts ensure that the experience of travel is primarily one whose aim is entertainment.

The barrage of films and documentaries with sensational overtones promoting the Nile as a place of mystery and adventure almost guarantees that it will not live up to that image. That women fiction writers have played a major role in creating the myth of the Nile, Egypt, and the Sahara as exotic and romantic has a certain irony. For my generation, it may have started with Agatha Christie's *Death on the Nile* (Christie 1937), but the writings of Elizabeth Peters (aka Barbara Mertz, a Ph.D. Egyptologist), who began in 1973 with a series of eighteen archeo-fictions, have continued it (Peters 1988, 2006). Her Victorian feminist Egyptologist, Amelia Peabody Emerson, appeals to an enormous audience of female romance readers. That appeal is also evident in the writings of other romance writers such as Victoria Holt (1973) and historical fiction writers such as Pauline Gedge (1999, 2006), who have used Egypt, the Nile, and the desert as their settings.

Even "serious" travel writers seem unable to escape the allure of the Nile and the desert as romantic escape. A travel essay entitled "One Night in the Sahara" (Jones 2003) recounts how the author became lost and was rescued by a handsome Wodaabe, a desert nomad, who led her gently by the hand back to her camp. In *Men of Salt: Crossing the Sahara on the Caravan of White Gold*, Benanav (2006) describes a region in the southwest Sahara where truckers, salt miners, and the caravaners make altruistic agreements and selfless sacrifices to promote the welfare of others. A full-page ads in the *New York Times Book Review* (April 23, 2006) reads, "I traveled the world and discovered iUniverse." The self-publishing press, iUniverse, touts the book, *Madam, Have You Ever Really Been*

Happy?: An Intimate Journey Through Africa and Asia, by Meg Noble Peterson (2005), who "after a divorce, a career, and raising five kids . . ." decides to escape and backpack alone around the world. Her descriptions of Egypt represent a continuum of writings that romanticize travel in the region.

Against this background of mass appeal through popularization of place, I found the Nile the most difficult river to write. Although I had few illusions that my journey would reveal any great truths or even any personal revelations, I had hoped that the river, with its compelling images of pyramids, pharaohs, sphinxes, hieroglyphics, and travel-poster scenes of dark camels and palms against a red sunset, would offer an uncommon beauty.

Travelers have been coming to Egypt for centuries, and a sizable body of literature has evolved on traveling in Egypt (Manley and Abdel-Hakim 2004). Nearly three thousand years ago, Herodotus was not the first to visit and write that the Nile was a corridor into the past. The allure of the Nile has attracted many throughout history. Not only the casual visitor but scholars as well seem bewitched by the place. The fact that at least fifty-eight U.S. colleges and universities (and possibly even twice that number worldwide) offer programs in Egyptology, and that there are thousands of Egyptologists in professional organizations, shows that Egypt is a discipline of scholarly interest. There is even a professional organization (the Association for the Study of Travel in Egypt and the Near East) whose aims are to promote the history of travel and travelers to this region. That this organization has a substantial membership, a conference every two years where fifty or more papers are presented, that it publishes a biennial bulletin and a dozen or more books, suggests that travel to and in Egypt is a legitimate field of study.

Today Egypt's economy is heavily dependent on the travel and tourism industry, with about 11 percent of its GDP attributed to tourism, although with terrorist bombings over the last five years these statistics fluctuate. When economic statistics are quoted, there is, of course, no distinction made between tourists and travelers. To my great disappointment, I was unable to experience the Nile as a serious traveler but only as a tourist. When guides are more interested in cell phones and techno-gadgets, when they are more concerned with their expensive shoes and stylish clothes, more concerned with company policy and programmed tours on a clock than with a genuine interest in their country's history and beauty, and when the companies they work for are big corporations, the experience can be disheartening. No matter how much effort you put into a journey, if those you depend upon as guides and facilitators are more interested in how much money they can extract from the visitor, the journey and the place become a tourist trap.

At the Cairo airport, among the confusion of waiting for luggage, a young Egyptian woman with two grumpy kids began a conversation that evolved into a litany of her personal possessions. She boasted of two homes, one in Alexandria and one on Long Island. Surprised at such blatant materialism, I was to learn that this was only the beginning of what seemed to be a pattern of preoccupation with the accumulation of wealth. From the guide who wanted money for retrieving a lost bag to the hotel clerk who wanted a tip for showing us to our room, to a maintenance man who wanted money for fixing a nonexistent problem with the door lock, the first day was an endless collision with people with their hands out.

The drive to Giza along the freeways of Cairo was our first glimpse of the beige city which stretched out like an endless sand castle. During our stay in the city, we visited the Khan el Khalili bazaar, the Egyptian Museum, the Great Pyramids, and the Sphinx, which after centuries of gazing upon the desert sands now stares at the red facade of a Pizza Hut and a Kentucky Fried Chicken. My encounters included guards at the pyramids who tried to hustle us into buying them cokes, getting disoriented in the bazaar as our guide sat drinking tea in a café while we wandered in confusion among the narrow streets, and seeing piles of mummies stacked up like logs as we raced through the museum on a one-hour time schedule.

The trip to Siwa oasis was supposed to be a pleasurable excursion into the desert. Between Cairo and El-Alamein, lulled by the monotony of a featureless landscape and too demanding a driving schedule, our driver fell asleep and rear-ended an eighteen wheeler. The truck was the only vehicle on the road for miles. The impact slammed me against the front windshield, and I hit glass like a loose missile. The car nearly flipped over, but the driver managed to keep it upright, and we whirled to a skidding stop. The cracked window resembled a giant snowflake with a hole in the middle the size of my head. My glasses were mangled, and I saw squigglies worming their way across my vision. Initially I felt nothing, but within the hour, my head began to throb and I felt a little queasy. Several cups of tea at the roadside cafe, while we waited for a new car and driver, dispelled the nausea, and by the time they arrived, I was ready to continue.

The new driver, with a companion to keep him awake, rambled on and on about how terrible that the car, a brand new company car, was so badly wrecked. What a shame the new car was so smashed up. And it didn't have any insurance because it was brand new. Finally, I said, "Well really, what about me? You can buy a new car, but I can't buy a new head." And he smiled and said, "Oh, you poor thing," and blew me a kiss. Then without skipping a beat, he asked if I had any daughters. When I answered, "No," he said, "Oh, that's too bad, I was looking for an American wife." At that point, I figured any attempts at meaningful conversation with the fellow would probably be futile.

The coastal road was a hundred miles of almost continuous tourist facilities, condominiums, and resorts. The Mediterranean coast looked like the Florida coast: development, development, and more development. I had hoped to see the UNESCO Omayad Biosphere Reserve along this stretch of road. Its 172,973 acres of protected habitat were established in 1981 to provide a home for several rare endemic species, most notably the shrubby rock rose, or sun rose (*Helianthemum sphaerocalyx*), and the endangered Egyptian tortoise

(*Testudo kleinmanni*). Of course, the tiny tortoise is near extinction because its habitat has been destroyed by the type of development we were seeing. As we sped along the condominium coast, there was nothing to indicate a reserve. When I asked the driver about it, he had no idea what I was talking about. It was late, we needed to get to Siwa before dark, and much to my disappointment, we did not have time to stop and search for the reserve or to explore the landscape for the rock rose or the tortoise.

Turning south at Mersa Matruh, all recognizable landmarks disappeared, and the highway became a narrow lane into the desert. For nearly two hundred miles, the vast empty sand stretched out, broken only by a couple of caravans of fifteen to twenty camels and an occasional oil rig. The caravans moved across the horizon in slow motion like a film running on low voltage, and one of the oil rigs had a huge plume of smoke billowing from it. About an hour into the desert, a sand storm came up, and even the minimalist landscape disappeared. The dust was so dense we could see nothing but beige static outside the window. It was like looking at a gravy-colored blank TV screen. For over an hour, we rode in a cloud of sand and dust. The driver never asked if we wanted to turn back, but if he had, I might have been tempted. The dust thinned occasionally, enough to allow him to see the sign posts marking the road, but mostly he was driving blind. I don't know how he managed to stay on the road but he did. Once in a while, a huge supply truck would suddenly loom from the dust to let us know we were still on the road.

Finally the storm abated, the air cleared somewhat, and the empty landscape reappeared. It felt like a scene from *The Sheltering Sky* (Bowles 1949). We arrived at the outskirts of Siwa at dusk, and by the time we reached the lodge, it was nearly dark. About seventeen kilometers from town, the Adrere Amellal (which means "white mountain" in the Siwi Berber language) lay nestled at the base of a salt mountain on a salt lake. In the twilight, the lodge resembled an adobe pueblo in a Martian landscape. It was as stunning as its Web site descriptions (EQI 2005). The Web site claimed the ecolodge was constructed and maintained using sound ecological techniques. This was why we had chosen it. In a traditional Siwan style with kershef, a mixture of sun-dried salt rock mixed with clay, it had been built by local craftsmen. The manager, a wide grumpy fellow named Soleiman met us and directed the all-male staff to take us to our rooms. The men in long white galabayas floated like ghosts across the sand as they led us with lighted lanterns to a second building. Up the winding staircase we climbed to a room with a massive wooden door and wooden latches. Everything in the room was of native wood, from the palm-trunk ceiling beams to the chairs, knobs, and shelves of olive wood. Candles provided the only light. There was no electricity in the lodge. Water was solar heated; wastewater was treated in settling tanks and recycled onto the agricultural fields.

Dinner, cooked in a traditional manner over an open fire, was delicious. We met four Canadian women with whom we would be touring Siwa the next day. They were the wives of wealthy landowners in the timber business, and they were traveling through Egypt on their own private tour. They were especially excited about the fact that Prince Charles and the Duchess of Cornwall would be coming to the lodge in several weeks, and they wanted to see the rooms where he and his entourage would be staying. After dinner as my traveling companion and I walked across the sands to our room, the salt lake glimmered in the moonlight, and the terrain, streaked with the shadows of palm trees and boulders, was breathtaking. But I was troubled by what I was learning about this ecolodge which seemed to be more of a luxury lodge.

The next morning we climbed aboard a jeep owned by Abdallah Baghi, our local Siwan guide, and began a tour of Siwa. Over the course of the tour, we learned that Mr. Baghi was a graduate of the University of Alexandria. A teacher and the head of the education department administering the school curriculum, he had traveled to Europe and the United States. He was especially proud of a 1999 award from the United Nations Race Against Poverty Program in recognition for his work in establishing Siwa House, an ethnological museum designed to preserve the oasis's heritage, and for helping to preserve the traditional arts of the region. A soft-spoken, articulate gentleman, his kindly manner was a refreshing change from the other guides we had encountered.

The places he took us to were interesting, but it was all rather programmed. When I set up the travel arrangements, I had requested and had been assured that we would spend our time in the desert, as I was interested in the plants of the region. But that did not occur. We toured Siwa and had only a few hours in the evening in the desert. Mr Baghi led us through the streets of the old fortress city of Shali. Abandoned when the rains of 1926 melted it, he cautioned us to be careful of our footing. Cleopatra's bath was a slimy thirty-foot-wide concrete pool containing spring water, and the oracle of Amum, being excavated by a group of German archeologists, was a temple ruin. It was the site where Alexander the Great supposedly learned of his impending death while conquering Egypt. Gabal Al-Mauta, or the mountain of the dead, was a necropolis of honeycombed tombs holding the mummies of the late pharaoh period and the Greco-Roman period. The view from the hilltop revealed Siwa to be an emerald island of about thirty-five square miles in a sea of khaki sand. Surrounded by salt lakes, the 220 natural freshwater springs and 1,600 artesian wells provided water for thousands of olive and date palm trees.

Siwa House, with a few rooms of clothing and domestic utensils, provided a glimpse of traditional Siwa culture. Finally we were led to the market in the town square which was a collection of mostly dingy stalls, although the food stands were colorful with their large variety of fruits, vegetables, spices, and dry goods. I expected to see beautiful embroidered items in the shops, as I had read about a project to help revive the traditional handicrafts of Siwa. Much like the American Navaho rug weavers, the needlework, done by elderly women, was fading. The project was to provide income for three hundred poor women of Siwa. Since the culture of Siwa demands that women remain in their homes, the project was designed to allow them to generate income at home. We did not see many women on the streets, and those few who were about wore milayahs that totally covered them. Nor did we see any exquisite embroidered clothing, jewelry, or accessories in the shops.

Two of the Canadian women wanted to take a donkey cart around the square, so they negotiated with a young man who had a donkey and a cart for hire. After a drudgingly slow ride around the square, they arrived back where they started. As the ladies wandered off, I heard shouts from the boy's father, who bellowed at him, belittling him for not charging the women more money.

The short time we spent in the Great Sand Sea was the most memorable. Mr. Baghi leaped from his jeep, deflated the tires, revved the engine, and off we glided into a land as desolate and barren as any I had ever seen. The Great Sand Sea, stretching about 400 kilometers to the south of Siwa, borders Libya. Because the Libyan border is only about 50 kilometers away, Mr. Baghi said that he had to be mindful not to get too close. Riding the dunes was like a giant roller coaster ride. With the exception of an occasional broom-straw shrub (*Calligonum comosum*), there was no plant life at all. Only at the oases were there a few grasses. At the first oasis, tall grasses (probably *Panicum turgidum* and *Phragmites* species) grew around the spring, and as we soaked in the warm water, I watched them bend and sway in the desert breeze. At the second water hole, a salt lake, Mr. Baghi built a fire to boil water, and we had tea. The vegetation consisted of several species of grasses and stunted thorny acacias; their stems were the source of our fire wood. The next stop was a great depression full of fossils. We strolled around the ancient sea bed searching for the perfect shell and intact sand dollar. It was a reminder that the Sahara was once a green savanna with lakes and forests (deVilliers and Hirtle 2002). A final stop was on a high dune ridge, where we watched the sun dip below the horizon and turn the sands the color of roasted beets.

Our last morning at breakfast, a conversation with a member of a newly arrived French group proved to be yet another encounter with the rich and privileged clientele which the lodge seemed to attract. An obvious returnee, she chattered on and on about the virtues of the lodge, making a point of telling us of another, less expensive lodge, the Shali Lodge, located nearer town. Perhaps we did not look wealthy enough to be able to afford the Adrere Amellal, and she was kindly offering us an alternative for future travel. In this class-conscious world, I supposed she felt she needed to make that clear.

When I returned home, I made an effort to learn more about the ecolodge, Adrere Amellal. The hundreds of Web sites promoting it ranged from a business article naming it "Hotel of the Week" to reports such as the one entitled "The Conde Nast Green List 2005: More than Just Luxury . . ." and a blurb in the University of Wisconsin alumni newsletter (summer 2006) profiling the owner. The story I was able to piece together from the articles revealed the following. In 1996, Mounir Neamatallah visited Siwa for the first time. On various Web sites, Mr. Neamatallah is identified as Egypt's leading environmentalist. He was educated in the United States with a Ph.D. in Environmental Health and Quality Management from Columbia University, and is the owner of the Cairo-based environmental consulting agency Environmental Quality International (EQI). EQI has 200 employees and claims that "all of EQI's activities aim to help the disadvantaged."

In 1997, funded by a program called the Environmental Sustainable Tourism Program, a joint effort of USAID and the Ford Foundation, Mr. Neamatallah built the luxury lodge, Adrere Amellal and the less luxurious Shali Lodge. USAID is the United States foreign aid agency, established by the Foreign Assistance Act of 1961, whose mission is to provide economic and humanitarian aid to countries in need. Its 2006 budget was $14.5 billion, and it is funded by U.S. taxpayers, so Mr. Neamatallah's lodges were paid for, in part, with my tax dollars.

EQI created the Siwa Women's Native Artisanship Development in 2001, financed by the International Finance Corporation (IFC) Grassroots Business Initiative (GBI). The project was to build two working facilities to train 350 women in traditional Siwan handicrafts and to establish links for the exportation of the women's artisan products. The International Finance Corporation (IFC) is the private-sector arm of the World Bank. This discovery was troublesome because the World Bank has a long history of funding projects that have resulted in deforestation, hydroelectric dams, displacement of the poor, and the degradation of natural environments. The former president of the World Bank, Paul Wolfowitz, is one of the leading architects of the Iraq war. The World Bank has often been criticized for policies that make the wealthy richer, and the poor, poorer (Goldman 2006).

In 2005, IFC again gave $1.9 million to EQI for "sustainable development." All the activities—the embroidery project, the building of the luxury ecolodge, the promoting of the Siwan culture and heritage, and the repairing of the old city of Shabli—are part of EQI's tourism development plan called the Siwa Sustainable Development Initiative. On EQI's Web site, the company claims that "600 Siwans have been gainfully employed," which represents about 3 percent of Siwa's 23,000 citizens.

Since 1996, EQI has acquired millions of dollars in funds from numerous organizations to develop Siwa. Using phrases like *sustainable development, sustainable tourism, eco-development, green development, heritage/cultural preservation, organic farming*, and other environmentally friendly terms, Siwa is being promoted as an eco-friendly place for the green-minded tourist. To the casual visitor, Siwa seems to be a charming, remote village that has changed little with time, a place whose ancient Berber culture has survived the empires of the Pharaohs, Greeks, Romans, Arabs, and the World Wars of Europeans and Americans. I wondered if it would survive the forces of modernity, commercialization, corporatization and the commodification of place. I wondered if it would survive development even when those activities were framed in the eco-friendly packaging of "sustainable development." I wondered why all the millions of dollars put into Siwa were not more visible, and what poor and disadvantaged people had been helped.

As an ecologist, my understanding of economics is not so-phisticated, and beyond basic food, water, shelter, and medical needs, I tend to view education as the most productive means of progress. I wonder if a culture that so restrictively limits its women from participating in public affairs is worth preserving in its entirety. Would not a more valuable direction of progress be to educate women to be strong, wise, and capable of negotiating through a complex and ever-changing technological world? I suspect that "sustainable development" may be a myth used to perpetuate a system that allows wealthy entrepreneurs, financiers, and managers of corporations to become even wealthier. I wonder what kind of future lies ahead when we spend our time and money destroying our natural environment and building artificial ones that promote tourism. While claims that tourism can be a power-ful conservation tool are common (Tsui 2006), I wonder if we are just being greenwashed.

It was not until we traveled to Alexandria to visit the new library that we encountered a less touristy environment. The road into the city cuts through marshland and tidal creeks, giving us our first view of the delta wetlands. Fishermen stood in tiny pirogues setting out nets among the tall stands of tasseled reeds (*Phragmites australis*). Stands of the salt-tolerant bulrush (*Typha capensis*) and needle rush (*Juncus maritimus*) also grew along the banks. The marshes of papyrus (*Cyperus papyrus*) that once covered the delta are gone. With the increased salinity, the papyrus has disappeared and more salt-tolerant species have taken over. There are no re-gions of the delta that remain undisturbed, and only a few pro-tected areas exist at all. Lake Burullus Protected Area west of Alexandria is the only preserve in the delta. The 113,668-acre pre-serve is the least disturbed of the delta wetlands, with reed-swamps

(*Phragmites* and *Typha* species) dominating in the north and mostly pondweed (*Potamogeton* species) in the south.

From aerial photographs, the Nile delta looks like a long-stemmed green blossom that grows smaller every year due to land loss. Currently, the delta is about the size of Massachusetts. Although the dams control the river's floods and allow farmers to grow three crops a year instead of one, as well as providing Egypt with about 15 percent of its hydroelectric power, they also prevent sediment deposition. Because of that, the wetlands are disappearing. The land is also becoming more saline as evaporation exceeds freshwater input, and the accumulated salt is not being washed away. The dams have altered the floodplain of the delta ecosystem dramatically in the last fifty years.

About 28 million people live in the delta in three large cities (Alexandria, Cairo, and Port Said), several large towns, and thousands of small villages. Along the two main distributaries, farms grow cotton, rice, onions, beans, wheat, corn, barley, and sugar. The fishing industry was drastically altered upon completion of the Aswan High Dam in 1970, with many species disappearing and those remaining on the decline. The sardine fishery, the bulk of the fishery harvest, declined 98 percent in the first few years but has recovered somewhat in recent years. The loss of freshwater from the Nile has increased salinity and altered the fishing of the entire Mediterranean Sea.

The World Wildlife Fund characterizes the ecology of the lower Nile as having 553 plants, including a hundred species of grasses, of which only eight are endemic. There are no lotus remaining—either the indigenous blue lily (*Nymphaea caerulea*) or the sacred lotus (*Nelumbo nucifera*) introduced by the Romans from southeast Asia. Not even invasives like water hyacinths, water lettuce, or Salvina are extensive. In short, the Nile delta is a heavily impacted region.

The Nile River is not much of a naturalist's paradise either. Chugging up and down the river, more than three hundred cruise boats carry several million tourists a year. Like mammoth floating hotels, they slurp up and down the river with their lazy cargo of sun-seekers. Our boat, the *Alexander the Great*, was an older one and actually looked like a real ship, but it was still a touristy hulk with forty passengers who had mainly come from the cold and damp of Europe to defrost in the dry, sunny climate of the Egyptian desert.

Along with a gazillion other tourists, we flew from Cairo to Luxor; the boats arrive and depart from Luxor and travel up and down the river to Aswan. The river is not heavily traveled from Cairo to Luxor, because it is only a few meters deep and too shallow for large ship navigation. Built on the site of the ancient city of Thebes, Luxor offered us the first of the great river temples. We walked among the pylons, pillars, and walls of hieroglyphics, dodging the throngs of tourists in the Karnak and Luxor temples. Located on the east bank of the river, the temples were easy walking distance from the boat, but the tombs in the Valley of the Kings were on the west bank far out into the desert. To get to the necropolis, we drove south along the Cairo-Aswan Highway, crossed the river, and made our way out into the arid hills. Passing through the lush green wheat fields speckled with plovers and crows, our guide claimed that the government had confiscated some of the farmland for road widening. Farmers whose livelihood depended on the land and whose status and identity were interwoven with land ownership were angry, and their attempts to stop this example of eminent domain were in the courts. Even in the Nile Valley, the most ancient of agricultural lands, conversion of land to uses other than agriculture seemed inevitable. Farmers all over the world seem to be losing farmland as economic forces swallow it up for other more profitable uses.

The Valley of the Kings was a stark graveyard. The lack of vegetation and the colorless gravel hills produced a monochromatic landscape. At the parking lot we were herded onto carts pulled by tractors to the tomb sites. Over sixty tombs lie in the Valley of the

Kings, and our tickets allowed us a walk-through of three of them. The tombs were barren except for the spectacularly painted and chiseled walls of hieroglyphics and scenes of royalty. In the Ramses VI, Tomb 9, row upon row of red and gold images of pharaohs adorned the ceiling like paper doll chains, cut and unfolded to make repeating figures, while boats ferried the procession of kings and their entourages across the river to the afterlife. In other chambers, colored hieroglyphics covered white stone walls. The columns of animals were a vertical zoo of quail chicks, vultures, owls, cranes, rabbits, snakes, and others. Some hieroglyphs depicted body parts: a mouth, an arm, feet, and hands. Some represented common items: twisted flax, a basket with a handle, a loaf of bread, folded cloth, a mat. Some were designs of natural things: water, the slope of a hill, a pool, a papyrus stem, or reeds.

A Google search gives dozens of Web sites that explain the more than two thousand symbols as a kind of alphabet. The story of Jean-Francois Champollion's discovery of the Rosetta stone and the deciphering of the hieroglyphics has been told by Meyerson (2004). Once again, it seems that even scholars (Meyerson is an Ellis Fellow at Columbia University) have been unable to escape Egypt's romantic appeal; the book has been described as a "florid adventure tale . . . as if drawn from a Hollywood film treatment" and a romance that "treats neither history nor linguistics with any degree of seriousness" (see Amazon.com, Publishers Weekly review of book).

Our guide claimed the hieroglyphics on the tomb walls were transcriptions of ancient religious texts from *The Book of the Dead* and other such funerary books. Spells, charms, and magical writings, they were designed to help the dead pass through the dangers of the underworld into a blissful afterlife. As I gazed on the inscribed walls, I thought how visually beautiful these texts were and how long they had survived. Then I thought of my own writing, how plain my alphabet and my words are compared to these. Yet I wondered if the need to make them came from the same desire. It seemed universal, this need to write our world and to create other worlds. Writing seems to be the way our imaginations and hopes take form, even a way to create hope.

At the Luxor Papyrus Institute—it was actually a shop—one of the artisans demonstrated the process of making paper. He cut the triangular stems of the papyrus into strips, overlapped and flattened them as he explained that the plant no longer grew wild but had to be purchased from plantations. He crossed other strips of the pulpy fiber to make a latticework and hammered them together with a wooden mallet. He explained that some of the first written records were cuneiform script on clay tablets, but when papyrus was invented, it revolutionized writing. Records could be kept on scrolls and preserved for long periods of time. The Dead Sea Scrolls, although many are on animal skins called vellum, represented only a few of the millions of fragments of ancient writing on papyrus located in museums and other facilities around the world. Papyrologists who studied such artifacts claim papyrus fragments include letters, novels, school notes, sales receipts, lease agreements, almost anything that might be recorded. Finally he blotted the sheet and said that when it dried it would be suitable for writing or painting. As a consummate salesman, he then showed us some of the paintings available for purchase.

The second day in Luxor, a sandstorm blew in from the west and filled the air with grit. We watched as the river disappeared in a light-brown fog. Then the dock disappeared and before long there was no scenery at all. Visibility was so reduced that the ship could not disembark, and we waited all afternoon for the storm to dissipate. Finally, late in the evening, the wind died down, the air cleared, and we were able to depart. A line of a dozen ships slowly backed away from the riverbanks and began the convoy downriver like a sluggish caravan of mastodons.

Like all rivers, the Nile begins small. The Blue Nile has its source in the Ethiopian highlands, the White Nile in Lake Victoria in Uganda, and the Atbarah River in Ethiopia. They merge in Sudan and for over 4,100 miles their waters flow north through

desert and finally empty into the Mediterranean Sea. The upper
Nile of the Sudan-Ethiopia region and the Uganda-Tanzania region has crocodiles, hippos, and other fauna and flora more typical
of an African river; the Egyptian or lower Nile is basically an agricultural canal.

While not sterile, the Nile has been impacted by centuries of
human intervention. For eight thousand years, humans have been
cultivating the banks of the river, and there are no vestiges of the
Nile Valley's original natural ecosystem left. What grows along the
riverbanks are cultivated crops and planted trees. Trees are either
ornamental or crop trees, and there are no native species. Predominantly the trees are palms, but others include drought-tolerant tama-
risk, acacia, eucalyptus, mimosa, jacaranda, cypress, and sycamores.
There are also orchards of citrus, fig, and mango.

Of the thousands of species of palms, eleven are native to
Africa and two grow abundantly along the Nile: the shorter
Y-trunk doum palm (*Hyphaene thebaica*) and the tall thin-trunk
date palm (*Phoenix dactylifera*). The doum palm stabilizes the
riverbanks and provides timber and an edible fruit. Its apple-sized
reddish-brown fruit tastes like gingerbread and contains antioxi-
dants, antifungals, antihypertensive, and cholesterol-lowering compounds. The fruits have a long history of consumption, having been
found in pharaohs' tombs and believed to be part of the ancients'
diet. Today a product called Doum powder, made by Sekem, the
organic food/herbal company, is sold in powder form to make a
gingery drink. The date palm also provides timber and dates. These
icons of the desert, with their tall, thin trunks and crowns of drap-
ing fronds, were the healthiest palm trees I had ever seen, and the
dates served at every meal were the sweetest I had ever eaten.

On the banks of the river, pumping stations lifted water up
onto the fields to irrigate the narrow strips of farmland. Each farm
averages about an acre. The river channel meandered, narrowed,
and expanded. Sometimes it was no wider than 150 meters and
other times it was half a mile wide. In some spots, the river banks

had no vegetation, and desert sand dunes hugged the shore. In other places, dark-brown mountains beyond the banks added height and texture to the tri-colored background. The river water was grey-green, the fields and palms a brilliant green, and the desert was the color of light-brown sugar. The landscape reminded me of a golf course, but it was not a land for casual recreation, except for the passing tourists. It was a well-maintained farm that feeds the people of Egypt.

Travelers do not come to the Nile to see a pristine natural ecosystem, although the birding was good in some places. I spent hours watching pied kingfishers hover above the water in search of fish, and many species of wading birds were common along the shore. Visitors come to see the remains of ancient civilizations. And so it was that we moved along the river for two days, stopping at the temples of Kom Ombo, Horus, Isis, and others, until they all began to blur together into a single image of crumbling stones.

In Edfu, a banner across the road from the Temple of Horus exhorted in large red English lettering, "Boycott all products from Denmark." When I asked our guide why the recent cartoon had caused such as a reaction, he proclaimed that all over the Islamic world it had caused great fury. He said the cartoon mocked his religion and that "no one should disrespect another person's religion." He was very articulate in protesting his opinion of disrespect shown towards his religion. His righteous outrage was actually the first sign of any sense of passion or interest from him. Throughout our travels, he had halfheartedly explained the sights in an indifferent manner that seemed programmed. He often recited the barebone facts quickly, and when asked a question often ignored it or answered with irritation. But when he spoke about the issue of his religion and the disrespect the recent cartoons had insinuated, it was as if he was finally speaking on a topic that he cared about. While I was glad to see some passion in the fellow, I wondered if he ever considered how I (and every other woman on earth) had lived our entire lifetimes with disrespect. I wondered if he ever thought about how women endured jokes that made fun of them, ads that used female bodies to sell cars and every imaginable product, behavior that relegated women to second-class citizenship, the endless violence against women, and that his own treatment of us had been rather shabby, and bordering on disrespect. I am often amazed at who gets to be the most legitimate victim in the seemingly never-ending conflict of who is most important. Yet beyond his hypocrisy, I recognized how that protest on the banner was an example of the written word giving voice to a universal desire. The words illuminated a need every human being possesses. The words proclaimed, "I want respect." And that is what writing does. It gives substance to ideas, to feelings, and to our needs. It transforms the personal into the universal. It gives us power to be less than invisible, to be heard.

At Aswan our cruise ended, and we flew back to Cairo along with the hordes of other tourists. I wondered, as I left the temples and tombs of the great valley, if the river served as a reminder of the importance of the written word. Of all that remains of the ancient civilizations, the writing seemed the most memorable. Perhaps the truth revealed by the Nile River was that we are creatures that must write our world.

Conclusion

The rivers I traveled were many and worldwide. Those in my own country were, of course, the most familiar, and I felt the most comfortable near them, but the others in foreign lands allowed for a sense of recognition as well. The question that arises from my journeys is what did these rivers offer? What have I learned in these travels? While each river provided the immediacy of place, each offered a particular awareness of certain environmental issues, and I was able to reflect on them as well as to present some natural history and engage in a bit of philosophizing. Rivers offer a way of understanding that perhaps

no other place can. They act as metaphors for human life and our
stories, life with all its commonalities and inconsistencies, rationali-
ties and irrationalities, paradoxes and congruities, ways of organiz-
ing and failures of organization.

The rivers I traveled offered me the plants and animals of our
world. Observing plants is not just for little old ladies, naturalists,
biologists, or gardeners. Everyone can relate at some level to plants.
The plants I observed—some unusual, some exotic, some rare and
endangered, some invasive, all interesting—were reminders of that
intimate connection we have with plants. They are our essential
companions on life's journey. Some of the rivers offered sightings
of rare and beautiful birds, and if we can believe the statistic that
46 million Americans are birders, we obviously relate to birds. Our
interest in birds may reflect a need to know that such creatures are
part of our world. And we need to connect with them. I believe this
applies to other animals as well, even insects. The animals I saw
along my river journeys were all wonders to behold.

Some rivers and their watersheds helped me understand the
urgency and necessity of protecting them. The old-world rivers,
especially the Yangtze and the Ganges, revealed the vulnerability of
rivers. The relationship of humans to these rivers is very old, because humans have been part of them for thousands of years. These
rivers suggested that no matter how much technology we may
develop, no matter how much spiritual reverence we have for the
river, if we do not keep our rivers safe and clean, we may be doomed
as a species.

The link between the exploitation of natural resources, the
status of women, poverty, and the corporate takeover of lands and
resources are presented with a subtlety that I hope allows for a
beginning awareness. These issues are certainly linked, and their
cause might be defined as the political power structures or the
personal characteristics of arrogance and greed. As an ecologist, I
tend to emphasis the biological and have not framed the issues in
especially strong political terms, but the issues are undeniably there.

Of all the specific themes in these stories, the overriding one that connects it all is that rivers are places of meaning. These writings are about how rivers give us the earth's treasures and how rivers connect us to each other and to our fellow species. Whether the river is in our immediate region or halfway around the world, there is something deeply familiar and connecting about a river. The river is both metaphorically and in reality our home. It is with this broader sense of connection that I have offered my reflections on a river time.

References

Adams, Jonathan S. 2006. *The Future of the Wild: Radical Conservation for a Crowded World*. Boston: Beacon Press.

Alley, Kelly D. 2002. *On the Banks of the Ganga: When Wastewater Meets a Sacred River*. Ann Arbor: University of Michigan Press.

Alter, Stephen. 2001. *Sacred Waters: A Pilgrimage Up the Ganges River to the Source of Hindu Culture*. New York: Harcourt.

Altringham, J. 1998. *Bats: Biology and Behavior*. Oxford: Oxford University Press.

Alves, Lise. 2006. "Middleman convicted in 2005 murder of U.S. born nun in Brazil." Catholic News Service. April 27, 2006. http://www.catholicnews.com/data/stories/cns.

Amazon.com. http://www.amazon.com/. Editorial Reviews of *The Linguist and the Emperor*. From *Publishers Weekly*.

Arana, Marie. 2006. *Cellophane*. New York: Dial Press.

ASLE. 2005. http://www.biblioserver.com/asle/

Attenborough, David. 1995. *The Private Lives of Plants*. London: BBC Books.

———. 1998. *The Life of Birds*. London: BBC Books.

Azuma, Hiroshi, L. B. Thien, and Kawano Shoichi. 1999. "Floral scents, leaf volatiles and thermogenic flowers in Magnoliaceae." *Plant Species Biology*. 14:121–127.

Babu, B. G. and M. Kannan. 2002. "Lightning Bugs." *Resonance*. 7:49–55.

Balick, Michael J., Elaine Elisabetsky, and Saraha A. Laird. (eds.) 1995. *Medicinal Resources of the Tropical Forest: Biodiversity and Its Importance to Human Health*. New York: Columbia University Press.

Barbour, Michael, John Evarts, and Marjorie Popper. (eds.) 2001. *Coast Redwoods: A Natural and Cultural History*. Los Olivos, CA: Cachuma Press.

Baskin, Yvonne. 2002. *A Plague of Rats and Rubbervines: The Growing Threat of Species Invasives*. Washington, D.C.: Shearwater Books/Island Press.

Bass, Rick. 1995. *The Lost Grizzlies: A Search for Survivors in the Wilderness of Colorado*. Boston: Houghton Mifflin.

Bat Conservation International. http://www.batcon.org

Bates, Henry. 1862. *The Naturalist on the River Amazon*. (1975 reprint edition). New York: Dover Publications.

Benanav, Michael. 2006. *Men of Salt: Crossing the Sahara on the Caravan of White Gold*. New York: The Lyons Press.

Berry, Wendell. 1977. *The Unsettling of America: Culture and Agriculture*. San Francisco: Sierra Club Books.

Bidartondo, M. I. 2005. "Tansley Review: The evolutionary ecology of myco-heterotrophy." *New Phytologist*. 167:335–352.

Bidartondo, M.I. and T. D. Bauns. 2001. "Extreme specificity in epiparasitic Monotropoideae (Ericaceae): Widespread phylogenetic and geographic structure." *Molecular Ecology*. 10:2285–2295.

———. 2002. "Fine-level mycorrhizal specificity in the Monotropoideae (Ericaceae): Specific for fungal species groups." *Molecular Ecology*. 11:557–569.

Bierregaard, Richard O., Jr., T. E. Lovejoy, Claude Gascon, Rita Mesquita. 2001. *Lessons from Amazonia: The Ecology and Conservation of a Fragmented Forest*. New Haven: Yale University Press.

Birchard, Bill. 2005. *Nature's Keepers: The Remarkable Story of How The Nature Conservancy Became the Largest Environmental Group in the World*. San Francisco: Jossey-Bass/Wiley Imprint.

Black, George. 2004. *The Trout Pool Paradox: The American Lives of Three Rivers*. New York: Houghton Mifflin.

Blake, Tupper Ansel, and Madeleine Graham Blake. 2000. *Balancing Water: Restoring the Klamath Basin*. Berkeley: University of California Press.

Bowles, Paul. 1949. *The Sheltering Sky*. London: John Lehmann.

Brewer, Richard. 2003. *Conservancy: The Land Trust Movement in America*. Lebanon, NH: University Press of New England.

Brox, Jane. 2004. *Clearing Land: Legacies of the American Farm*. New York: North Point Press.

———. 1999. *Five Thousand Days Like This One: An American Family History*. Boston: Thorndike Press.

———. 1995. *Here and Nowhere Else: Late Seasons of a Farm and Its Family*. Boston: Beacon Press.

Buck, Pearl. 1932. *The Good Earth*. London: Methuen and Co. Ltd.

Bulter, Linda. 2004. *Yangtze Remembered: The River Beneath the Lake*. Palo Alto: Stanford University Press.

Burdick, Allan. 2005. *Out of Eden: An Odyssey of Ecological Invasion*. New York: Farrar, Straus, and Giroux.

Busch, Robert H. 2001. *Grizzly Almanac*. New York: The Lyons Press.

Cahill, Tim. 1997. "A Darkness on the River." *Pass the Butterworms: Remote Journeys Oddly Remembered*. New York: Vintage Press.

Camel, Nancy. 2006. *The Nature of Things at Lake Martin: Exploring the Wonders of Cypress Island Preserve*. Lafayette, LA: Acadian House Publishing.

Campbell, David G. 2005. *A Land of Ghosts: The Braided Lives of People and the Forest in Far Western Amazonia*. New York: Houghton Mifflin.

Camuto, Christopher. 2000. *Another Country: Journeying Towards the Cherokee Mountains*. Athens: University of Georgia Press.

Carter, Jimmy. 1994. *An Outdoor Journal*. Fayetteville: University of Arkansas Press.

Cashwell, Peter. 2003. *The Verb "to bird:" Sightings of an Avid Birder*. Philadelphia: Paul Dry Books.

Castner, James L. 2000. *Explorama's Amazon: A Journey Through the Rainforest of Peru*. Gainesville: Feline Press.

Chagnon, Napoleon. 1983. *Yanomamo: The Fierce People*. (3rd edition) New York: Holt, Rinehart, Winston.

Champney, Lizzie. 1885. *Three Vassar Girls in South America*. Boston: Estes and Lauriat.

Chetham, Deirdre. 2002. *Before the Deluge: The Vanishing World of the Yangtze's Three Gorges*. New York: Palgrave Macmillan.

Christie, Agatha. 1937. *Death on the Nile*. London: Avon Books.

Clark, Leonard. 1953. *The Rivers Ran East*. New York: Funk & Wagnalls Company.

Coastkeepers. 2006. http://www.coastkeepers.org.

Cocker, Mark. 2003. *Birders: Tales of a Tribe*. New York: Grove Press.

Coleman, Kate. 2005. *The Secret Wars of Judi Bari: A Car Bomb, the Fight for the Redwoods, and the End of Earth First!* San Francisco: Encounter Books.

Conroy, Pat. 1986. *The Prince of Tides*. New York: Houghton Mifflin.

Corbett, Jim. 1946. *Man-eaters of Kumaon*. Oxford: Oxford University Press.

Cornell, Judith. 2001. *Amma: Healing the Heart of the World*. New York: Harper Collins.

Cronin, John, and Robert Kennedy, Jr. 1997. *The Riverkeepers: Two Activists Fight to Reclaim Our Environment as a Basic Human Right*. New York: Scribner.

Dallmeyer, Dorinda G. 2004. "Epilogue: Why We Write." *Elemental South: An Anthology of Southern Nature Writers*. Athens: University Georgia Press.

Davis, Wade. 1997. *One River: Explorations and Discoveries in the Amazon Rain Forest*. New York: Simon and Schuster.

Deblieu, Jan. 1991. *Meant to Be Wild: The Struggle to Save Endangered Species Through Captive Breeding*. Golden, CO: Fulcrum Publishers.

———. 1998. *Wind: How the Flow of Air Has Shaped Life, Myth, and the Land*. New York: Houghton Mifflin.

De Villiers, Marq and Sheila Hirtle. 2002. *Sahara: A Natural History*. New York: Walker and Company.

Dinerstein, Eric. 2005. *Tigerland and Other Unintended Destinations*. Washington. D.C.: Shearwater Books/Island Press.

Dobrin, Sidney and Christopher Keller (eds.). 2005. *Writing Environments*. Albany: State University of New York Press.

Dowie, Mark. 2006. "Conservation Refugees." Tim Folder (ed.), *The Best American Science and Nature Writing—2006*. New York: Houghton Mifflin.

Downie, Andrew. 2005. "Death of nun shows peril of Amazon activism." *Christian Science Monitor*. February 16, 2005. http://www.csmonitor.com/2005/0216/p05s01-woam.html

Earthwatch. 1997. "China Three Gorges Dam-Eco-Boon or Cesspool?" http://www2.cnn.com/EARTH/9711/04/china.dam.reut

Economy, Elizabeth C. 2004. *The River Runs Black: The Environmental Challenge to China's Future*. Ithaca: Cornell University Press.

Emanations. 1998. http://labs.plantbio.cornell.edu/cbl/Pubs/Emanation.html

EQI. 2005. www.eqi.com.eg/.

Fermor, Patrick Leigh. 2005 (reprint edition). *Between the Woods and the Water: On Foot to Constantinople from the Hook of Holland: The Middle Danube to the Iron Gates*. New York: The New York Review of Books.

Findley, James. 1995. *Bats: A Community Perspective*. (reprint edition) Cambridge: Cambridge University Press.

Fitzgerald, Carol. 2001. *The Rivers of America: A Descriptive Bibliography*. 2 vols. New Castle, DE: Oak Knoll Press.

Food Time Line History Notes. http://www.foodtimeline.org/foodfaq.html# applesauce.

Forsman, Eric, E. C. Meslow, and H. M. Wright. 1983. "Distribution and biology of the spotted owl in Oregon." *Wildl. Monograph*. 87:1–64

Forsman, Eric. 2002. *Natal and Breeding Dispersal of Northern Spotted Owls*. Wildlife Society Monograph no. 149. Bethesda: Wildlife Society.

Frankenberg, Dick. 1997. *The Nature of North Carolina's Southern Coast: Barrier Islands, Coastal Waters and Wetlands*. Chapel Hill: University of North Carolina Press.

Gedge, Pauline. 1999. *Lord of the Lands: The Hippopotomas Marsh*. New York: Penguin.

———. 2006. *The Horus Road*. New York: Penguin.

Ghosh, Amitav. 2004. *The Hungry Tide*. Delphi: Ravi Dayal.

Gilbert, Elizabeth. 2006. *Eat, Pray, Love: One Woman's Search for Everything Across Italy, India, and Indonesia*. New York: Viking.

Goldman, Michael. 2006. *Imperial Nature: The World Bank and Struggles for Social Justice in the Age of Globalization*. New Haven: Yale University Press.

Gordon, Deborah. 1999. *Ants at Work: How an Insect Society Is Organized*. New York: The Free Press.

Gotwald, William H., Jr. 1995. *Army Ants: The Biology of Social Predation*. Ithaca: Comstock Publishing Associates/Cornell University Press.

Gould, Lewis. 1988. *Lady Bird Johnson: Our Environmental First Lady*. Lawrence: University of Kansas Press.

Goulding, Michael, R. Barthem, and E. Ferreira. 2003. *The Smithsonian Atlas of the Amazon*. Washington D.C.: Smithsonian Books.

Halicka, H. D., B. Ardelt, G. Juan, A. Miltelman, S. Chen, F. Tarqanos, and Z. Darzynkiewicz. 1997. "Apoptosis and cell cycle effects induced by extracts of the Chinese herbal preparation PC-SPES." *International Journal of Oncology*. 11:437–448.

Hall, James. 2006. "Wall Street Reaps Big Bucks From the Wind." May 1, 2006. http://capitalhillcoffeehouse.com/more.php?id=91_0_1_18_M.

Harper, F. 1958. *Travels of William Bartram*. Naturalist's edition. New Haven: Yale University Press.

Harris, David. 1997. *The Last Stand: The War Between Wall Street and Main Street Over California's Ancient Redwood*. San Francisco: Sierra Club Books.

Hawkins, R. E. (ed.) 1978. *Jim Corbett's India: Stories Selected by R. E. Hawkins*. Oxford: Oxford University Press.

Helvarg, David. 1997. *The War Against the Greens: The Wise Use Movement, The New Right, and Anti-Environmental Violence*. San Francisco: Sierra Club Books.

Herring, Hal. 2005. "Room to Maneuver: Can a New Partnership Between the Military and the Conservationists Prove That What's Good for the Black Hawk is also Good for the Black Bear?" *The Nature Conservancy Magazine* 54:21–29.

Hersey, John. 1956. *A Single Pebble*. New York: Alfred Knopf.

Herzog, Werner. 1982. *Fitzcarraldo*. DVD release. 1999. Anchor Bay Studio.

Hess, Paul. 2005. "News and Notes: Intelligence in Corvids." *Birding*. 37:43–44.

Hessler, Peter. 2002. *Rivertown: Two Years on the Yangtze River*. New York: Harper Collins.

————. 2006. *Oracle Bones: A Journey Between China's Past and Present*. New York: Harper Collins.

Hilderbrand, R. H., A. C. Watts, and A. M. Randle. 2005. "The myths of restoration ecology." Ecology and Society (1)19. http://www.ecologyand society.org/vol10/iss1/art19.

Hirt, Paul. W. 1994. *A Conspiracy of Optimism: Managing the National Forests Since World War Two*. Lincoln: University of Nebraska Press.

Hitt, Jack. 2006. "13 Ways of Looking at an Ivory-Billed Woodpecker." *The New York Times Magazine*. (May 7).

Hobshouse, Henry. 2003. *Seeds of Wealth: Four Plants That Made Men Rich*. New York: Shoemaker & Hoard.

Hodges, Elizabeth. 1999. *What the River Means*. Pittsburgh: Duquesne University Press.

Holldobler, Bert and E. O. Wilson. 1990. *The Ants*. Cambridge: Belknap Press/ Harvard University Press.

Holmes, Hannah. 2005. *Suburban Safari: A Year on the Lawn*. New York: Bloomsbury.

Holt, Victoria. 1973. *Curse of the Kings*. New York: Doubleday.

Honey, Martha. 1999. *Ecotourism and Sustainable Development: Who Owns Paradise*. Washington, D.C.: Island Press.

Hoyt, Eric. 1996. *The Earth Dwellers: Adventures in the Land of Ants.* New York: Simon & Schuster.

Hurd, Barbara. 2003. *Entering the Stone: On Caves and Feeling Through the Dark.* Boston: Houghton Mifflin.

International Crane Foundation. 2003. "The Siberian Crane Shangri-La-Poyang Lake China." http://www.savingcranes.org/Travel/tr-china.asp

International Ecotourism Society. http://www.ecotourism.org.

Jaramillo, Alvaro and Peter Burke. 1999. *New World Blackbirds: The Icterids.* Princeton: Princeton University Press.

Jones, Amanda. 2003. "One Night in the Sahara." George Don (ed.), *The Kindness of Strangers.* Melbourne, Australia: Lonely Planet Publications.

Jones, D. L. 1988. *Native Orchids of Australia.* Reed Books. New South Wales: Frenchs Forest.

Kane, Harnett T. 1943. *The Bayous of Louisiana.* New York: William Morrow.

Kaul, Veenu, A. K. Koul, and M. C. Sharma. 2000. "The Underground Flower." *Current Science* 78: 39–44.

Keiter, Robert B. 2003. *Keeping Faith with Nature: Ecosystems, Democracy and American's Public Lands.* New Haven: Yale University Press.

Kiss, Ida Miro. 2000. "The Blond is Dead: the Tisza River Disaster." Central Europe Review. http://www.ce-review.org/00/7/kiss7.html.

Knauth, Stephen. 1999. *The River I Know You By.* "Testimony." New York: Four Way Books. http://www.fourwaybbooks.com/books/knauth/knauth2.html.

Koeppel, Dan. 2005. *To See Every Bird on Earth: A Father, a Son and a Lifetime Obsession.* New York: Hudson Street Press.

Koh, Lian Pin, Robert R. Dunn, Navjot S. Sodhi, Robert K. Colwell, Hesther C. Proctor, and Vincent S. Smith. 2004. "Species Co-extinctions and the Biodiversity Crisis." *Science.* 305:1632–1634.

Komanoff, Charles. 2006. "Whither Wind? A Journey Through the Heated Debate Over Wind Power." *Orion.* Sept/Oct. http://www.orionsociety. org/ pages/om/06-5om/Komanoff.html.

Kricher, John. 1997. *A Neotropical Companion: An Introduction to the Animals, Plants and Ecosystem of the New World.* 2nd edition. Princeton: Princeton University Press.

Kurta, Allen, Armando Rodriquez-Duran, Michael R. Williq, and Michael R. Gannon. 2005. *Bats of Puerto Rico: An Island Focus and a Caribbean Perspective.* Lubbock: Texas Tech University Press.

Lanier, Sidney. 1884. *Poems of Sidney Lanier: Edited by His Wife: Electronic Edition.* http://docsouth.unc.edu/southlit/lanier1/lanier.html.

Lauck, Joanne Elizabeth. 2002. (revised edition). *The Voice of the Infinite in the Small: Revisioning the Insect-Human Connection.* Boston: Shambhala Press.

Leaf, Sue. 2004. *Potato City: Nature, History and Community in the Age of Sprawl.* Minneapolis: Borealis Book.

Leland, John. 2005. *Aliens in the Backyard: Plant and Animal Imports into America.* Columbia: University of South Carolina Press.

Levi-Strauss, Claude. 1955. *Tristes Tropiques*. (1989 edition). London: Pan Books.

Levine, Mark. 2002. "And Old Views Shall Be Replaced by New." Hal Espen (ed.), *Outside 25: Classic Tales and New Voices from the Frontiers of Adventure*. New York: Norton and Company.

Lindburg, Donald, and Karen Baragona. (eds.) 2004. *Giant Pandas: Biology and Conservation*. Berkeley: University of California Press.

Little, Charles E. 1995. *The Dying of the Trees: The Pandemic in America's Forest*. New York: Penguin.

Lockwood, C. C. 1982. *Atchafalaya: America's Largest River Basin Swamp*. Baton Rouge: Beauregard Press.

Longfellow, Henry Wadsworth. 1922. *Evangeline*. New York: Macmillian Company.

Lopez, Barry. 1978. *Of Wolves and Men*. New York: Scribner.

Louisiana State Flower. 2006. http://www.netstate.com/states/symb/flowers/la_magnolia.htm.

Lutz, Dick. 1999. *Hidden Amazon: The Greatest Voyage in Natural History*. Salem, OR: Dimi Press.

Lynd, Mitch. 2006. "Great Moments in Apple History." http://www.hort.purdue.edu/newcrop/maia/history.html.

MacCreagh, Gordon. 1961. *White Waters and Black*. Garden City: Doubleday and Company.

Macdonald, Sarah. 2002. *Holy Cow: An Indian Adventure*. New York: Bantam Books.

Mackinnon, Karen Phillipps, and Fen-Qi He. 2000. *A Field Guide to the Birds of China*. Oxford: Oxford University Press.

Maclean, John N. 2003. *Fire and Ashes: On the Front Lines of America's Wildfires*. New York: Henry Holt and Co.

Maloof, Joan. 2005. *Teaching the Trees: Lessons from the Forest*. Athens: University of Georgia Press.

Mancall, Peter C. (ed.). 1996. *Land of Rivers: America in Word and Image*. Ithaca: Cornell University Press.

Manley, Deborah and Sahar Abdel-Hakim. 2004. *Traveling Through Egypt: From 450 BC to the Twentieth Century*. Cairo: The American University in Cairo Press.

Matthiessen, Peter. 2001. *The Birds of Heaven: Travels With Cranes*. New York: Farrar, Straus and Giroux.

McDonald, Hamish. 2004. "Class, Religion Spark Riots across China." http://www.theage.com.au/articles/2004/11/02/1099362142005.html?from= storylhs).

McLaughlin, Andrew. 1993. *Regarding Nature: Industrialism and Deep Ecology*. Albany: State University of New York Press.

McKibben, Bill. 2003. "Speaking Up for the Environment." Marie Arana (ed.), *The Writing Life: Writers on How They Think and Write: A Collection from the Washington Post Book World*. Washington, D.C: Public Affairs.

McPhee, John. 1989. "Atchafalaya." *The Control of Nature*. New York: Farrar, Straus and Giroux.

Means, Bruce. 1991. "Florida's Steepheads: Unique Canyonlands." *Florida Wildlife*. 45:75–92.

Mehta, Gita. 1994. *Karma Cola: Marketing the Mystic East*. (reprint edition) New York: Vintage.

———. 1994. *A River Sutra*. (reprint edition) New York: Vintage.

Mehta, Suketa. 2004. *Maximum City: Bombay Lost and Found*. New York: Knopf.

Menzies, Gavin. 2002. *1421: The Year China Discovered America*. New York: Harper Collins.

Merrill, Arch. 1986. *Southern Tier (Arch Merrill's New York)*. (reprint edition). Interlaken, NY: Heart of the Lakes Publishing.

Meyerson, Daniel. 2004. *The Linguist and the Emperor: Napoleon and Champollion's Quest to Decipher the Rosetta Stone*. New York: Random House.

Millard, Candice. 2005. *The River of Doubt: Roosevelt's Darkest Journey*. New York: Doubleday.

Mishra, Pankaj. 1995. *Butter Chicken in Ludhiana: Travels in Small Town India*. New York: Penguin Books.

Montgomery, Sy. 2000. *Journey of The Pink Dolphins: An Amazon Quest*. New York: Simon and Schuster.

Moore, Kathleen Dean. 1995. "The Willamette." Kathleen Dean Moore, *Riverwalking: Reflections on Moving Water*. New York: Harcourt Brace & Company.

———. 1999. *Holdfast: At Home in the Natural World*. New York: The Lyons Press.

———. 2004. *The Pine Island Paradox: Making Connection in a Disconnected World*. Minneapolis: Milkweed Editions.

Morgan City. 2005. http://www. cityofmc.com/history.html.

Morrow, Susan Brind. 2004. *Wolves and Honey: A Hidden History of the Natural World*. New York: Houghton Mifflin.

Murray, John A. (ed.) 1998. *The River Reader*. New York: The Lyons Press.

———. 1992. *The Great Bear: Contemporary Writings on the Grizzly*. Anchorage: Alaska Northwest Books.

Naipaul, V. S. 1991. *India: A Million Mutinies Now*. New York: Penguin Books.

National Wild and Scenic Rivers System. http://www.nps.gov/rivers/about.html.

The Nature Conservancy. http://www.tnc.org.

Navdanya. http://www.navdanya.org.

Neely, Jack. 2004. "Tellico Dam Revisited." Metro Pulse Online. http://www.metro pulse.com/dir_zine/dir_2004/1450/t_cover.html.

Newby, Eric. 1966. (U.S. editions 1998). *Slowly Down the Ganges*. Melbourne: Lonely Planet Publications.

Newsom, Deanna and Daphne Hewitt. 2005. "The Global Impacts of SmartWood Certification: Final report." http://www/rainforest-allianace.org/programs/forestry/perspectives/documents/sw%5Fimpacts.pdf.

Nijhuis, Michelle. 2006. "Selling the Wind." *Audubon*. (Sept–Oct). 108(5): 52–93.

Noss, Reed. 1999 (ed.). *The Redwood Forest: History, Ecology and Conservation of the Coastal Redwoods*. Washington, D.C.: Island Press.

Obmascik, Mark. 2004. *The Big Year: A Tale of Man, Nature and Fowl Obsession*. New York: Free Press.

Osborne, Jane. 2003. *I'd Rather Be Birding*. College Station: Texas A & M University Press.

Osborne, Lawrence. 2006. *The Naked Tourist: In Search of Adventure and Beauty in the Age of the Airport Mall*. New York: North Point.

Padel, Ruth. 2005. *Tigers in Red Weather*. New York: Little, Brown.

Palmer, Tim. 2004. *Endangered Rivers and the Conservation Movement*. 2nd edition. Lanham, MD: Rowman and Littlefield.

Parker, Andrew. 2005. *Seven Deadly Colours: The Genius of Nature's Palette and How it Eluded Darwin*. London: Free Press/Simon and Schuster.

Peacock, Doug. 1996. *Grizzly Years: In Search of the American Wilderness*. New York: Owl Books.

Peters, Elizabeth. 1988. *Crocodile on the Sandbank*. (re-issue edition). New York: Mysterious Press.

———. 2006. *Tomb of the Golden Bird*. New York: William Morrow.

Peterson, Meg Noble. 2005. *Madam, Have You Really Ever Been Happy?: An Intimate Journey Through Africa and Asia*. iUniverse. www.iuniverse.com/whyiuniverse/different-journey/meg.htm.

Phillips, Dana. 2003. *The Truth of Ecology: Nature, Culture, and Literature in America*. Oxford: Oxford University Press.

Plotkin, Mark. 1994. *Tales of a Shaman's Apprentice: An Ethnobotanist Searches for New Medicines in the Amazon Rainforest*. New York: Penguin Press.

Pollan, Michael. 2006. *The Omnivore's Dilemma: A Natural History of Four Meals*. New York: Penguin Press.

Postel, Sandra and Brian Richter. 2003. *Rivers for Life: Managing Water for People and Nature*. Washington D.C.: Island Press.

Prance, G. T. and T. E. Lovejoy. 1985. *Amazonia*. New York: Pergamon Press.

Pressley, Nelson. 2006. "The Churning Mind of Deepa Mehta: In the Fight to Film 'Water,' Director Confronted Elemental Forces in India." The Washington Post (May 7) http://www.washiungtonpost.com/wp-dyn/content/article/2006/05/05/AR2006050500405.htm.

Preston, Richard. 2006. "Climbing the Redwoods." Atul Gawande (ed.), *The Best American Science Writing—2006*. New York: Harper Perennial.

———. 2007. *The Wild Trees: A Story of Passion and Daring*. New York: Random House.

Pyne, Stephen J. 2001. *Fire: A Brief History*. Seattle: University of Washington Press.

———. 2004. *Tending Fire: Coping With America's Wildfires*. Washington, D.C.: Island Press.

Qing, Dai. 1997. *The River Dragon Has Come: The Three Gorges Dam and the Fate of China's Yangtze River and Its People*. Armonk, NY: M. E. Sharpe, Inc.

Quammen, David. 1998. "Planets of Weeds: Tallying the Loss of Earth's Animals and Plants." *Harper's Magazine*. 297 (1781):57–69.

———. 2003. *Monster of God: The Man-eating Predators in the Jungles of the History and Mind*. New York: W. W. Norton and Company.

Raffles, Hugh. 2002. *In Amazonia: A Natural History*. Princeton: Princeton University Press.

Ray, Janisse. 2005. *Pinhook: Finding Wholeness in a Fragmented Land*. White River Junction, VT: Chelsea Green Publishing.

Reece, Erik. 2006. *Lost Mountain: A Year in the Vanishing Wilderness*. New York: Riverhead/Penguin.

Ritchie, Mark, 1995. *The Spirit of the Rainforest: A Yanomamo Shaman's Story*. New York: Island Lake Press.

Roberts, Gregory David. 2003. *Shantaram*. New York: St. Martin's Griffin.

Roberts, M. R. and L. Zhu. 2002. "Early response of the herbaceous layer to harvesting in a mixed coniferous deciduous forest in New Brunswick, Canada." *Forest Ecology and Management*. 155:17–31.

Rollins, James. 2003. *Amazonia: A Novel*. New York: Avon Press.

Roosevelt, T. D. 1914. *Through the Brazilian Wilderness*. London: J. Murray.

Rothenberg, David. 2005. *Why Birds Sing: A Journey into the Mystery of Birdsong*. New York: Basic Books.

Roy, Arundhati. 1999. "The Great Common Good." *The Cost of Living*. New York: Modern Library.

Rushdie, Salman. 2004. "The Firebird's Nest." Nadine Gordimer (ed.), *Telling Tales*. New York: Picador.

Salzman, Mark. 1984. *Iron and Silk*. New York: Vintage Books.

Sarnoff, Paul. 1965. *Russell Sage: The Money King, The Man Who Banked the Tycoons*. New York: Ivan Obolensky.

Schaller, George B. 1994. *Last Panda*. Chicago: University of Chicago Press.

Schrepfer, Susan. 1983. *The Fight to Save the Redwoods: A History of Environmental Reform, 1917–1978*. Madison: University of Wisconsin Press.

Schultes, R. E. and Siri von Reis. (eds.). 1995. *Ethnobotany: Evolution of a Discipline*. Portland: Timber Press.

See, Lisa. 2003. *Dragon Bones*. New York: Random House.

Shiva, Vandana. 2002. *Water Wars: Privatization, Pollution, and Profit*. Cambridge: South End Press.

Shiva, Vandana and Radha Holla-Bhar. 1997. "Piracy by Patent: The Case of the Neem Tree." Jerry Mander and Edward Goldsmith (eds.), *The Case Against the Global Economy: And for a Turn Toward the Local*. San Francisco: Sierra Club Books.

Shoumatoff, Alex. 1978. *The Rivers Amazon*. San Francisco: Sierra Club Books.

Simberloff, D. and Betsy Von Holle. 1999. "Positive interactions of non-indigenous species: invasional meltdown." *Biological Invasions*. 1:21–32.

Simpson, Bland and Ann Cary Simpson. 1997. *Into the Sound Country: A Carolinian's Coastal Plain*. Chapel Hill: University of North Carolina Press.

Singh, Ashok, V. S. Reddy, and J. S. Singh. 1995. "Analysis of woody vegetation of Corbett National Park." Plant Ecology. 120 (no 1):69–79.

Slater, Candace. 2002. *Entangled Edens: Visions of the Amazon*. Berkeley: University of California Press.

Slater, Candace. (ed.). 2003. *In Search of the Rain Forest: New Ecologies for the Twenty First Century*. Durham: Duke University Press.

Smith, Nigel. 1999. The *Amazon River Forest: A Natural History of Plants, Animals and People*. Oxford: Oxford University Press.

Snyder, Gary. 1994. *Coming into the Watershed*. New York: Pantheon.

———.1995. *A Place in Space: Ethics, Aesthetics, and Watersheds*. Washington, D.C.: Counterpoint.

Solnit, Rebecca. 2007. *Storming the Gates of Paradise: Landscapes for Politics*. Berkeley: University of California Press.

Spruce, Richard. 1908. *Notes of a Botanist on the Amazon and Andes*. London: MacMillian and Company.

Stavinoha, William B. and Neera Satsangi. 2005. "Status of *Ganoderma lucidium* in United States: *Ganoderma lucidium* as an anti-inflamation agent." http://www.kyotan.com/lecture/lectures/Lecture4.html.

Stille, Alexander. 2003. "The Ganges' Next Life." Richard Miller and Kurt Spellmeyer (eds.), *The New Humanities Reader*. New York: Houghton Mifflin.

Stone, Roger R. 1985. *Dreams of Amazonia*. New York: Penguin Press.

St. Petersburg Times. 2001. "Yellow River Dam is Vetoed." (June 20).

Suzuki, David and Wayne Grady. 2004. *Tree: A Life Story*. Vancouver: Greystone Books.

Swift, Earl. 2005. "Can they be stopped?" *Parade Magazine*. (May 22).

Taber, Stephen. 1998. *The World of the Harvest Ants*. College Station: Texas A & M University Press.

Tan, Amy. 1989. *The Joy Luck Club*. New York: Putnam.

Terzani, Tiziano. 1997. *A Fortune-Teller Told Me: Earthbound Travels in the Far East*. New York: Harmony Books.

Theroux, Paul. 1983. *Sailing Through China*. Norwich: Michael Russell Publishing Ltd.

———. 1988. *Riding the Iron Rooster: By Train Through China*. New York: Random House.

Thomas, J. A., M. G. Telfer, D. B. Roy, C. D. Preston, J. J. D. Greenwood, J. Asher, R. Fox, R. T. Clarke, and J. H. Lawton. 2004. "Comparative Losses of British Butterflies, Birds, and Plants and the Global Extinction Crisis." *Science* 303:1879–1881.

Thoreau, Henry David. 1983. *The Maine Woods*. (reprint edition). Princeton: Princeton University Press.

Tidwell, Michael. 1996. *Amazon Stranger: A Rainforest Chief Battles Big Oil*. New York: The Lyons Press.

———. 2003. *Bayou Farewell: The Rich Life and Tragic Death of Louisiana's Cajun Coast*. New York: Pantheon Books.

Tierney, Peter. 2000. *Darkness in El Dorado: How Scientists and Journalists Devastated the Amazon*. New York: Norton.

Trust for Public Lands http://www.tpl.org.

Tsui, X. 2006. "Journeys: Ecotourism; Traveling the World to Help Save It." The *New York Times*. (December 17).

Twain, Mark. 1883. *Life on the Mississippi*. Boston: James R. Osgood and Company.

U.S. Water News Online. 2002. "Proposal Would Dam Florida Yellow River to Provide Water Sources." http://www.uswaternews.com/archives/arcpolicy/2prowou7.html.

Vaughan, Elizabeth. 2004. "The Appalachicola Bluffs and Ravines Preserve in Northern Florida: A Long Leaf Pine and Wiregrass Restoration Project." http://www.hort.agri.umn.edu/h5015/01papers/vaughan.htm.

Vileisis, Ann. 1997. *Discovering the Unknown Landscape: A History of America's Wetlands*. Washington, D.C.: Island Press.

Wagenknecht, Louise. 1998. "And the Salmon Sing." John Murray (ed.). *The River Reader*. New York: The Lyons Press.

Walker, Sue and Dennis Holt. 2004. *In the Realm of Rivers: Alabama's Mobile-Tensaw Delta*. Montgomery: New South Books.

Wallace, Alfred Russell. 1895. *A Narrative of Travels on the Amazon and Rio Negro*. London: Ward, Lock, and Bowden, Ltd.

Wallace, David. 2003. *The Klamath Knot: Explorations of Myth and Evolution, Twentieth Anniversary Edition*. Berkeley: University of California Press.

Wallace, Scott. 2003. "Into the Amazon." *National Geographic Magazine*. 204:2–27.

Watkins, Nan. 2002. *East Towards Dawn: A Woman's Solo Journey Around the World*. New York: Seal Press.

Weber, Thomas. 1988. *Hugging the Trees: The Story of the Chipko Movement*. New Delhi: Viking.

Weidensaul, Scott. 2005. *Return to Wild America: A Yearlong Search for the Continent's Natural Soul*. New York: North Point Press.

Weinstein, Barbara. 1983. *The Amazon Rubber Boom, 1850–1920*. Palo Alto: Stanford University Press.

Wheeler, William B. and Michael J. McDonald. 1986. *TVA and the Tellico Dam: A Bureaucratic Crisis in Post Industrial America*. Knoxville: University of Tennessee Press.

Wilderness Classroom Organization. 2006. Pacaya Samiria National Reserve: Macaws, monkeys, and illegal loggers. May 2, 2006. http://news. mongabay. com/2006/0502-wc11.html.

The Wilderness Society. 2006. http://www.wilderness.org/NewsRoom? Statement/20060209.cfm.

Williams, Michael. 2002. *Deforesting the Earth: From Prehistory to Global Crisis; An Abridgment*. Chicago: University of Chicago Press.

Wilson, Don E. and Merlin D. Tuttle. 1997. *Bats in Question: The Smithsonian Answer Book*. Washington D.C.: Smithsonian Books.

Winchester, Simon. 1996. *The River at the Center of the World: A Journey up the Yangtze, and Back in Chinese Time*. New York: Henry Holt and Company.

Wohl, Ellen. 2004. *Disconnected Rivers: Linking Rivers to Landscape*. New Haven: Yale University Press.

Woodard, Colin. 2004. "Protecting the heartwood: By saving the core of the forest, scientists hope to safeguard the whole." *The Nature Conservancy Magazine.* 54 (1):42–50.

———. 2006. "The Sale of the Century." *The Nature Conservancy Magazine.* 56 (3):20–25.

World Heritage Sites. UNEP WCMC. 2005. http://sea.unep-wcmp.org/site/pa/ 0081v.htm.

World Wildlife Fund. 2003. News: Illegal Harvest Bigleaf Mahogany Confiscated in Peru. Feb 05. 2003. http://www.panda.org/news_facts/newsroom/news/ index.cfm?uNewsID=5706.

———. 2004. Nile Delta Flooded Savanna (PA0904) http://www.worldwildlife. org/wildworld/profiles/terrestrial/pa/pa0904_full.html.

Xinhua News Agency. 2005. "Beijing steel giant may suspend production to curb pollution during 2008 Olympics." March 8. http://www.china.org.cn/english/ environment/124732.htm.

Xinran. 2002. *The Good Women of China: Hidden Voices.* New York: Anchor Books.

Yamashita, Karen. T. 1990. *Through the Arc of the Rainforest.* Minneapolis: Coffee House Press.

Young, Emma. 2002. "Yangtze River polluted at dangerous levels." *New Scientist.* (Print edit) 13:20. (January 17) http://www.newscientist.com/article.ns?id= dn1802&print=true

Ziewitz, Kathryn and June Wiaz. 2004. *Green Empire: The St. Joe Company and the Remaking of Florida's Panhandle.* Gainesville: University Press of Florida.

Zuckerman. Larry. 1998. *The Potato: How the Humble Spud Rescued the Western World.* New York: Farrar, Straus and Giroux.

Index